WHY

SCIENTISTS

ACCEPT

EVOLUTION

Robert T. Clark and James D. Bales

WHY SCIENTISTS ACCEPT EVOLUTION

by
Robert T. Clark
and
James D. Bales

"Many a false theory gets crystallized by time and absorbed into the body of scientific doctrine through lack of adequate criticism when it is formulated." (Professor William Diller Matthew, *Climate and Evolution*, 2nd edition, p. 159)

BAKER BOOK HOUSE
Grand Rapids, Michigan

Library of Congress Catalog Card Number: 66-18305

Standard Book Number: 8010-2310-6

First printing, May 1966
Second printing, October 1967
Third printing, November 1970

PREFACE

The doctrine of evolution is regarded today as an established fact of science. David J. Merrell maintained that anyone who considered with an open mind the evidence for evolution could scarcely remain unconvinced that evolution had taken place.[1] "...no rational mind can question the invincible nature of the evolutionary case."[2] Thus wrote H. G. Wells. "...at the present time," others have written, "no unprejudiced student can possibly reject what the authors of *The Science of Life* have termed 'the incontrovertible fact of evolution,' and no responsible scientist does reject it."[3]

So widely accepted is the doctrine of evolution that it is received by each oncoming generation for the simple reason that each generation finds that evolution is a part of the scientific world outlook in which it is reared. It is assumed that the battle concerning the truth or falsity of evolution was adequately waged in the nineteenth century, and that evolution won the day because scientific confirmation was set forth by the evolutionists.

To suggest that there is a possibility that evolution has not been scientifically proved is to raise a question about one's knowledge or one's integrity, as has already been pointed out. But is it possible that in the name of science itself the question of the truth or falsity of evolution should be seriously studied by our generation? Richard B. Goldschmidt, an outstanding evolutionist, prefaced one of his articles with a statement from a

---

[1] *Evolution and Genetics*, New York: Holt, Rinehart and Winston, 1962, p. vii.

[2] H. G. Wells, *Mind at the End of Its Tether*, New York: Didier Publishers, 660 Madison Ave., 1946, pp. 29-30. This was one of Wells' last books. Concerning such a claim the reader is referred to Douglas Dewar's answer to it in *More Difficulties of the Evolution Theory*, London: Thynne and Co., Ltd., 1938. Also see his final work on evolution, *The Transformist Illusion*, Murfreesboro, Tennessee: DeHoff Publishing Company, 1957.

[3] C. W. Young *et al.*, *The Human Organism and the World of Life*, New York: Harper and Brothers, 1938, p. 293.

5

letter that Orville Wright wrote on June 7, 1903. Wright said: "...but if we all worked on the assumption that what is accepted as true is really true, there would be little hope of advance."[4]

However, it will take more than a statement by Wright to convince evolutionists that the question may need to be re-examined. It would surely be granted that it would not be unscientific to reopen the question, if it could be shown that it was not the weight of scientific evidence which led in the nineteenth century to the acceptance of evolution. What if evolution was accepted because of the determination to rule out the real possibility of creation by God and to place in its stead a naturalistic explanation of life's origin and manifold forms? If it was decided that all must be explained naturally, then obviously one would have to accept some hypothesis of evolution regardless of whether or not it was scientifically established. Can it be that evolution was not proved by scientific evidence but that it was rather the inevitable result of an *a priori* decision, as expressed in this century by Julian S. Huxley, that: "Modern science must rule out special creation or divine guidance."[5]

In determining whether it was an *a priori* decision, the authors will not advance opinions or speculations of their own. Instead, they will consider the problem from the historical standpoint, as a question of this nature must be considered if it is to be answered rightly. The problem is: Was there in the mind of Darwin, and others of the nineteenth century, a dogma that made it impossible for them to criticize adequately the hypothesis of evolution before it became, in the words of William Diller Matthew, absorbed into the body of scientific doctrine?

The aim of this treatise, therefore, is not to delve into the arguments pro or con for the hypothesis of evolution. It is to

---

[4] "Evolution, as Viewed by One Geneticist," *American Scientist,* Vol. 40, 1952, p. 84.

[5] *Evolution: The Modern Synthesis,* New York: Harper and Brothers, 1943, p. 457.

deal with the question as to the real reason why it was accepted in the nineteenth century and then passed on to the twentieth century. In the course of this study it will be shown that at least some evolutionists displayed more frankness in their letters than they did in their scientific works.

CONTENTS

CHAPTER I

JAMES HUTTON (1726-1797) AND UNIFORMITARIANISM

The doctrine of uniformity is the foundation of the modern acceptance of the various hypotheses of evolution. As T. H. Huxley put it: "The doctrine of evolution in biology is the necessary result of the logical application of the principles of uniformitarianism to the phenomena of life. Darwin is the natural successor of Hutton and Lyell, and the 'Origin of Species' the logical sequence of the 'Principles of Geology.'"[1] One may hold, of course, to the uniformity of the laws of nature without being committed to the extreme position that present-day causes explain all past events. However, uniformity as viewed by Darwin, Huxley, and others implied evolution.

James Hutton is generally regarded as the father of uniformitarianism, and he may also "fairly be called one of the founders of modern geology."[2] Hutton stated the doctrine of uniformity as follows: "Not only are no powers to be employed that are not natural to the globe, no action to be admitted of except those of which we know the principle, and no extra-ordinary events to be alleged in order to explain a common appearance, the powers of nature are not to be employed in order to destroy the very object of those powers; we are not to make nature act in violation to that order which we actually observe, and in subversion of that end which is to be perceived in the system

---

[1] *Darwiniana*, New York: D. Appleton and Co., 1896, p. 232.
[2] Frank Dawson Adams, *The Birth and Development of the Geological Sciences*, Baltimore: The Williams and Wilkins Co., 1938, p. 239.

of created things. In whatever manner, therefore, we are to employ the great agents, fire and water, for producing those things which appear, it ought to be in such a way as is consistent with the propagation of plants and the life of animals upon the surface of the earth. Chaos and confusion are not to be introduced into the order of nature, because certain things appear to our partial views as being in some disorder. Nor are we to proceed in feigning causes, when those seem insufficient which occur in our experience."[3]

Since Hutton's hypothesis involved "the exclusion of all causes not supposed to belong to the present order of Nature,"[4] Hutton had to explain all past events in terms of present-day causes. If present-day causes do not seem to be sufficient to explain all things, one must postulate vast periods of time in the past in order to give these causes time to produce these physical changes. "In order to produce the present continents, the destruction of a former vegetable world was necessary; consequently, the production of our present continents must have required a time which is indefinite."[5] His conclusion concerning time was that "we find no vestige of a beginning,—no prospect of an end."[6]

Hutton limited his hypothesis of uniformity to the world of physical things. He did not seem to rule out the idea of God, and he realized that his theory dealt only with the present order of nature. Thus he wrote, concerning the system of rivers, that it "is then to be considered as an object of design; and, in this design, we may perceive either wisdom, so far as the ends and

[3] Quoted from "Theory of the Earth," in Kirtley F. Mather and Shirley L. Mason, *A Source Book in Geology,* New York: McGraw-Hill Book Co., Inc., 1939, p. 95.

[4] Lyell, as quoted in *More Letters of Charles Darwin,* New York: D. Appleton and Co., 1903, Vol. II, p. 149, footnote.

[5] James Hutton, "Theory of the Earth," *Transactions of the Royal Society,* Edinburgh, 1785, Vol. I, p. 301. Quoted by Harold W. Clark, *The New Diluvianism,* Angwin, California: Science Publications, 1946, p. 5.

[6] Hutton, *op. cit.,* p. 304.

means are properly adapted, or benevolence, so far as that system is contrived for the benefit of beings who are capable of suffering pain and pleasure or of judging good and evil."[7]

Let us now trace the influence of this hypothesis on the geologist Lyell, whose successful propagation of it not only influenced the framing of hypotheses of evolution, but also prepared the way for its widespread acceptance.

---

[7] *Theory of the Earth,* Vol. II, p. 566. Thanks to Robert E. D. Clark, Cambridge University, for sending me this quotation along with a microfilm of Hutton's *Theory of the Earth.* For another statement that indicates Hutton's belief in God see Mather and Mason, *op. cit.,* p. 95.

CHAPTER II

SIR CHARLES LYELL,
THE UNIFORMITARIAN GEOLOGIST

As Charles Darwin and T. H. Huxley acknowledged, it was Sir Charles Lyell who played a large part in the preparation of the scientific world, and of many laymen, for the hypotheses of evolution. Sir Charles Lyell was the scientist who did the most to convince the scientific world that the doctrine of uniformity was a basic law. His extensive writings, and his lectures in England, Europe, and America did much to popularize the theory. It is, therefore, important for us to consider his relationship to the doctrine of uniformity and to the hypotheses of evolution.[1]

Huxley on Lyell's Influence

"From 1830 onwards for more than forty years Lyell's 'Principles of Geology' was one of the most widely read scientific books in this country...."[2] Lyell's uniformitarianism, when taken to its logical conclusion, "postulates evolution as much in the

---

[1] In our research concerning Lyell material will often be drawn from the *Life, Letters, and Journals of Sir Charles Lyell.* This two-volume set, written by his sister-in-law, Mrs. Lyell, was issued by John Murray, in London, in 1881. Unable to find a set in the libraries to which the authors had access, and unable to purchase a set although it was advertised for in the United States and England, the authors were deeply grateful to Professor George McCready Price (who passed on in 1963), of Loma Linda, California, for the generous loan of his set.

[2] Huxley's *Life and Letters,* London: Macmillan & Co., Ltd., 1903, Vol. III, p. 18.

organic as in the inorganic world."[3] "I cannot but believe that Lyell, for others, as for myself [Huxley], was the chief agent for smoothing the road for Darwin."[4] In writing to Lyell he urged on him that evolution was implied in his doctrine of uniformity.[5] Without uniformity the "whole theory crumbles to pieces."[6]

Lyell, Hutton and Uniformity

Although Lyell had once accepted Buckland's "catastrophical" theory, he had abandoned it for uniformity before he wrote the first volume of the *Principles of Geology*.[7] At the age of seventeen in 1816 he had matriculated at Exeter College, Oxford. There he attended the geology lectures of Dr. Buckland. An interest in geology, and in the antiquity of the earth, already existed. "Bakewell's 'Geology,' which he found in his father's library, was the first book which gave him an idea of the existence of such a science as geology, and something said in it about the antiquity of the earth excited his imagination so much that he was well prepared to take interest in the lectures of Dr. Buckland, Professor of Geology at Oxford...."[8]

It was not long until he could speak of "my bias towards a leading doctrine of the Huttonian hypothesis."[9] "In 1827 Mr. Lyell wrote in the 'Quarterly Review' an article on Scrope's 'Geology of Central France,' in which he showed how entirely he had imbibed the opinions of Playfair and Hutton, and considered that all geological monuments were to be interpreted

---

[3] Huxley in the *Life and Letters of Charles Darwin*, New York: D. Appleton & Co., 1898, Vol. I, p. 544.

[4] *Ibid.*, Vol. I, pp. 543-544. "And most contemporaries agreed that Lyell was the most important single influence preparing the way for Darwin." Gertrude Himmelfarb, *Darwin and the Darwinian Revolution*, London: Chatto & Windus, 1959, p. 153.

[5] *Life and Letters of T. H. Huxley*, Vol. I, p. 252.

[6] *Life and Letters of Charles Darwin*, Vol. I, p. 553.

[7] Lyell, *op. cit.*, Vol. II, pp. 6, 7.

[8] *Ibid.*, Vol. I, p. 32.

[9] *Ibid.*, Vol. II, p. 4.

by reference to aqueous and igneous causes in action in the ordinary course of nature."[10]

In a letter of January 15, 1829 he viewed uniformity to be of extreme importance for geology. In fact, his book on geology was designed to furnish illustrations of this principle. Of the book he wrote: "It will not pretend to give even an abstract of all that is known in geology, but it will endeavour to establish the *principle of reasoning* in the science; and all my geology will come in as illustration of my views of those principles, and as evidence strengthening the system necessarily arising out of the admission of such principles, which, as you know, are neither more nor less than that *no causes whatever* have from the earliest time to which we can look back, to the present, ever acted but those *now acting;* and that they never acted with different degrees of energy from that which they now exert."[11]

Although he said that he had tried to find such, he did not believe that there was any evidence to overthrow uniformity. "As a staunch advocate for absolute uniformity in the order of Nature, I have tried in all my travels to persuade myself that the evidence was inconclusive, but in vain."[12]

The doctrine of uniformity, however, did not mean to him that the same things happened over and over again. "My notion," he wrote, "of uniformity in the existing causes of change always implied that they must for ever produce an endless variety of effects, both in the animate and inanimate world."[13]

This doctrine of uniformity he advocated so long and vigorously that one of the tributes to him, written after his death, spoke of his works as follows: "His leading lesson was a belief in the uniformity of the laws of nature: a belief which led him to argue that by studying the changes which are being wrought upon the surface of the earth by the silent action of forces now in operation, we put ourselves in possession of a key to the

---

[10] *Ibid.,* Vol. I, p. 160.
[11] *Ibid.,* Vol. I, p. 234.
[12] *Ibid.,* Vol. I, p. 260. Feb. 3, 1830.
[13] *Ibid.,* Vol. II, pp. 2-3. March 7, 1837.

interpretation of those ancient records which it is the special business of the geologist to decipher. Sir Charles, indeed, developed with singular success the great truths which were first enunciated by Dr. Hutton of Edinburgh, and eloquently illustrated by his friend Professor Playfair. Hutton died in 1797, and it is curious to note that the same year which witnessed his death gave birth to one who was destined to expound his doctrines with such force of argument as to carry them successfully against all opposition, and establish them as fundamental principles of the science."[14]

Lyell a Believer in God

Although a firm believer in uniformity Lyell did not, therefore, conclude that the universe was self-explained and self-contained. He believed that God, the Presiding Mind, exists.[15] In a letter to the Duke of Argyll, September 19, 1868, he wrote: "I cannot believe that Darwin or Wallace can mean to dispense with that mind of which you speak as directing the forces of nature. They in fact admit that we know nothing of the power which gives rise to variation in form, colour, structure, or instinct."[16]

To Charles Darwin, May 5, 1869, he said: "I quite agree with you that Wallace's sketch of natural selection is admirable. I wrote to tell him so.... I reminded him that as to the origin of man's intellectual and moral nature I had allowed in my first edition that its introduction was a real innovation, interrupting the uniform course of the causation previously at work on the earth. I was, therefore, not opposed to his idea, that the Supreme Intelligence might possibly direct variation in a way analogous to that in which even the limited powers of man might guide it in selection, as in the case of the breeder and horticulturist. In other words, as I feel that progressive development or evolution cannot be entirely explained by natural selection, I rather hail Wallace's suggestion that there may be a

---

[14] *Ibid.,* Vol. II, p. 470.
[15] *Ibid.,* Vol. I, p. 380.
[16] *Ibid.,* Vol. II, p. 432.

Supreme Will and Power which may not abdicate its functions of interference, but may guide the forces and laws of nature."[17]

A Beginning Probable

On June 14, 1830, in writing to Poulett Scrope who was about to review the *Principles of Geology,* Lyell stated that it was probable that there had been a beginning, but that there were no signs of it. His universal extension of uniformity always led him to conclude that whatever looked like a beginning was due to his ignorance of "all possible effects of existing causes" rather than actually being the beginning.

"Probably there was a beginning—it is a metaphysical question, worthy of a theologian—probably there will be an end. Species, as you say, have begun and ended—but the analogy is faint and distant. Perhaps it is an analogy, but all I say is, there are, as Hutton said, 'no signs of a beginning, no prospect of an end.' "[18]

A Creation

And yet as a "strong theist"[19] Lyell did not believe that evolution did away with the need for creation. Thus he told Darwin on March 11, 1863, that: "I think the old 'creation' is almost as much required as ever, but of course it takes a new form if Lamarck's views improved by yours are adopted."[20]

Against Moses

Lyell's acceptance of God and of the need for some creation

---

[17] *Ibid.,* Vol. II, p. 442. Lyell also believed in free-will. He also thought that death might not end all. Not so very long before his death he wrote to Miss F. P. Cobbe, July 20, 1873, that "for as I cannot help being convinced that I have the power of exerting free-will, however great a mystery the possibility of this may be, so the continuance of a spiritual life may be true, however inexplicable or incapable of proof." Vol. II, p. 452.

[18] *Ibid.,* Vol. I, pp. 269-270.

[19] As Charles Darwin called him, *Life and Letters of Charles Darwin,* Vol. I, p. 59.

[20] Lyell, *op. cit.,* Vol. II, p. 363.

did not mean that he accepted the Bible. This certainly would have been inconsistent with his dogma of uniformity. Science needed to be freed from Moses, he contended. In April, 1829, he spoke of driving certain men "out of the Mosaic record."[21] On June 14, 1830, he said of some that "they see at last the mischief and scandal brought on them by Mosaic systems."[22] The Mosaic deluge, especially, had been an incubus to the science of geology, or so he claimed August 29, 1831.[23]

Evolutionary Ideas in Lyell's Writings

In a letter to his father, February 7, 1829, Lyell expressed his antagonism to the idea of a simultaneous creation of various species. He thought that geology would help determine whether or not they came gradually. "I am now convinced that geology is destined to throw upon this curious branch of inquiry, and to receive from it in return, much light, and by their mutual aid we shall very soon solve the grand problem, whether the various living organic species came into being gradually and singly in isolated spots, or centres of creation, or in various places at once, and all at the same time. The latter cannot, I am already persuaded, be maintained."[24] "It is not the beginning I look for, but proofs of a *progressive* state of existence in the globe, the probability of which is *proved* by the analogy of changes in organic life."[25] "The changes in organic life, which I intend to be more generally entertaining than the inorganic, and more new, must be deferred to Vol. II."[26]

These ideas may have been the result of Lamarck's influence. In a letter of March 2, 1827 he stated that he had read, but had not accepted, Lamarck. And yet, even in the letter there are indications that the hypothesis of evolution must have had some influence on him. "I devoured Lamarck *en voyage,* as you did

---

21 *Ibid.,* Vol. I, p. 253; cp. 256.
22 *Ibid.,* Vol. I, p. 268; cp. 240.
23 *Ibid.,* Vol. I, p. 328.
24 *Ibid.,* Vol. I, pp. 245-246.
25 *Ibid.,* Vol. I, p. 270.
26 *Ibid.,* Vol. I, p. 271.

Sismondi, and with equal pleasure. His theories delighted me more than any novel I ever read, and much in the same way, for they address themselves to the imagination, at least of geologists who know the mighty inferences which would be deducible were they established by observations. But though I admire even his flights, and feel none of the *odium theologicum* which some modern writers in this country have visited him with, I confess I read him rather as I hear an advocate on the wrong side, to know what can be made of the case in good hands. I am glad he has been courageous enough and logical enough to admit that his argument, if pushed as far as it must go, if worth anything, would prove that men may have come from the Ourang-Outang. But after all, what changes species may really undergo! How impossible will it be to distinguish and lay down a line, beyond which some of the so-called extinct species have never passed into recent ones. That the earth is quite as old as he supposes, has long been my creed, and I will try before six months are over to convert the readers of the Quarterly to that of Heterodox opinion."[27]

Thus Lyell was an evolutionist at heart long before Darwin propounded his hypothesis. In fact, it was more or less implied in his *Principles* which Darwin read on the *Beagle*. Huxley quoted the above statement from Lyell and maintained it revealed that evolution was an esoteric doctrine with Lyell, and that he did not plainly commit himself to it because of the fear of offending certain persons and of arousing prejudices against himself.[28]

On November 23, 1868, Lyell wrote to Professor Haeckel concerning references which Haeckel had made in his *History of Creation* to Lyell's work. "I am therefore obliged to you for pointing out how clearly I advocated a law of continuity even in the organic world, so far as possible without adopting Lamarck's theory of transmutation. I believe that mine was the first work (published in January 1832) in which any attempt

27 *Ibid.*, Vol. I, p. 168.
28 *Life and Letters of Charles Darwin*, Vol. I, pp. 545-546.

had been made to prove that while the causes now in action continue to produce unceasing variations in the climate and physical geography of the globe, and endless migration of species, there must be a perpetual dying out of animals and plants, not suddenly and by whole groups at once, but one after another. I contended that this succession of species was now going on, and always had been; that there was a constant struggle for existence, as De Candolle had pointed out, and that in the battle for life some were always increasing their numbers at the expense of others, some advancing, others becoming exterminated.

"But while I taught that as often as certain forms of animals and plants disappeared, for reasons quite intelligible to us, others took their place by virtue of a causation which was beyond our comprehension; it remained for Darwin to accumulate proof that there is ño break between the incoming and outgoing species, that they are the work of evolution, and not of special creation."[29]

Lyell expressed a belief, in 1841, that progressive development would result in a creature superior to man. "We do not meet with many people of colour so far north, but without being an anti-abolitionist, I am sometimes half inclined to believe, that when the geological time arrives according to the system of progressive development for a being as much transcending the white man in intellect as the Caucasian race excels the chimpanzee...."[30]

Lyell's fundamental position thus involved the idea of new species. These new species were produced by the following agencies: "I allude to, first, the adequacy of known causes as parts of one continuous progression to produce mechanical effects resembling in kind and magnitude those which we have to account for; secondly, the changes of climate; thirdly, the changes from one set of animal and vegetable species to another."[31]

---

[29] Lyell, *op. cit.*, Vol. II, p. 436.
[30] *Ibid.*, Vol. II, p. 55. To his father-in-law, Lockport (near Niagara), August 26, 1841.
[31] *Ibid.*, Vol. II, p. 5.

Preparing People for Darwin

Not only were there evolutionary ideas in Lyell, but he recognized that these and uniformity helped prepare the minds of many for Darwin's doctrine. On November 23, 1868, he so expressed himself to Haeckel. "Most of the zoologists forget that anything was written between the time of Lamarck and the publication of our friend's 'Origin of Species.' "[32] "I had certainly prepared the way in this country, in six editions of my work before the 'Vestiges of Creation' appeared in 1842, for the reception of Darwin's gradual and insensible evolution of species, and I am very glad that you noticed this. . . ."[33]

Although he had not been able to go all the way with Darwin, Lyell told Joseph Hooker, March 9, 1863, that he influenced many to accept Darwin. "Darwin . . . seems much disappointed that I do not go farther with him, or do not speak out more. I can only say that I have spoken out to the full extent of my present convictions, and even beyond my state of *feeling* as to man's unbroken descent from the brutes, and I find I am half converting not a few who were in arms against Darwin, and are even now against Huxley."[34] "However, I plead guilty to going farther in my reasoning towards transmutation than in my sentiments and imagination, and perhaps for that very reason I shall lead more people on to Darwin and you, than one who, being born later. . . ."[35] To Charles Darwin, March 11, 1863, he expressed it as follows: "But you ought to be satisfied, as I shall bring hundreds towards you, who if I treated the matter more dogmatically would have rebelled."[36]

Beliefs Concealed

Henshaw Ward claimed that Lyell, as a matter of policy, kept some of his ideas from the readers and left them to be inferred. Lyell thought that new species were taking the place

---

[32] *Ibid.*, Vol. II, p. 436.
[33] *Ibid.*, Vol. II, pp. 436-437.
[34] *Ibid.*, Vol. II, p. 361.
[35] *Ibid.*, Vol. II, pp. 361-362.
[36] *Ibid.*, Vol. II, pp. 363-364.

of old ones that had been destroyed, but he was not always definite about it. "He kept it almost invisible, because he could not refer to any particle of a fact which would give any indication of how a species could be 'created' by natural law. He wanted his readers to gather such an idea, but wanted to avoid the scientific odium of proposing it as a part of science."[37] "It is therefore probable that Lyell believed his 'creation' of species was brought about by some quite unknown operation of natural law. It is certain that Lyell chose, as a matter of policy, to conceal his belief and to let readers infer that a 'creation' was a miraculous interposition of the hand of God. Judd considers that Lyell was justified in this policy...."[38] Justified or not, it was a characteristic of Lyell.

This effort to conceal some of his views was likely due to his desire to be thought well of and thus to avoid that which would bring him into reproach. While a boy he had hunted bugs, butterflies and such like. This practice was looked down on and his friends and family ridiculed him. The effect on him is described in his own words. "The disrepute in which my hobby was held had a considerable effect on my character, for I was very sensitive of the good opinion of others, and therefore followed it up almost by stealth; so that although I never confessed to myself that I was wrong, but always reasoned myself into a belief that the generality of people were too stupid to comprehend the interest of such pursuits, yet I got too much in the habit of avoiding being seen, as if I was ashamed of what I did."[39]

This same characteristic was found in his writings for many years. At times, however, he rationalized it to the place that he thought that it was not only right, but praiseworthy, and that the opposite course in this matter would have been wrong.

---

[37] *Charles Darwin and the Theory of Evolution,* New York: The New Home Library, 1943, p. 436.

[38] *Ibid.,* pp. 436-437. See also Gertrude Himmelfarb, *Darwin and the Darwinian Revolution,* pp. 153-160, 169.

[39] Lyell, *op. cit.,* Vol. I, p. 17.

Thus in a letter dated December 29, 1827, he said: "I marvel less at Dr. ....'s anticipations (as I supposed them) in geological speculation, now that I observe he followed Hutton, and cites him. I think he ran unnecessarily counter to the feelings and prejudices of the age. This is not courage or manliness in the cause of Truth, nor does it promote its progress. It is an unfeeling disregard for the weakness of human nature, for as it is our nature (for what reason Heaven knows), but *it is* constitutional in our minds, to feel a morbid sensibility on matters of religious faith. I conceive that the same right feeling which guards us from outraging too violently the sentiments of our neighbors in the ordinary concerns of the world and its customs should direct us still more so in this."[40]

May 11, 1830 he wrote: "But I assure you that I have been so cautious, that two friends tell me that I shall *only* offend the ultras."[41]

June 14, 1830 he further explained the purpose of his approach in saying that "If we don't irritate, which I fear that we may (though mere history), we shall carry all with us. If you don't triumph over them, but compliment the liberality and candour of the present age, the bishops and enlightened saints will join us in despising both the ancient and modern physico-theologians.... If I have said more than some will like, yet I give you my word that full *half* of my history and comments was cut out, and even many facts; because either I, or Stokes, or Broderip, felt that it was anticipating twenty or thirty years of the march of honest feeling to declare it undisguisedly."[42]

"I conceived the idea five or six years ago, that if ever the Mosaic geology could be set down without giving offence, it would be in an historical sketch, and you must abstract mine, in order to have as little to say as possible yourself. Let them feel it, and point the moral."[43]

---

[40] *Ibid.*, Vol. I, pp. 173-174.
[41] *Ibid.*, Vol. I, p. 267.
[42] *Ibid.*, Vol. I, p. 271. Compare Gertrude Himmelfarb, *Darwin and the Darwinian Revolution*, p. 320.
[43] *Ibid.*, Vol. I, p. 271.

On June 1, 1836, in a letter to Sir J. W. Herschel he expressed his belief in the origination of new species, but he more or less inferred it in his books so as not to arouse opposition. "In regard to the origination of new species, I am very glad to find that you think it probable that it may be carried on through the intervention of intermediate causes. I left this rather to be inferred, not thinking it worth while to offend a certain class of persons by embodying in words what would only be a speculation."[44]

"When I first came to the notion, which I never saw expressed elsewhere, though I have no doubt it had all been thought out before, of a succession of extinction of species, and the creation of new ones, going on perpetually now, and through an indefinite period of the past, and to continue for ages to come, all in accommodation to the changes which must continue in the inanimate and habitable earth, the idea struck me as the grandest which I had ever conceived, so far as regards the attributes of the Presiding Mind."[45]

On March 7, 1837, he explained why he did it this way. "In regard to this last subject ['the changes from one set of animal and vegetable species to another,' J.D.B.], as well as to change of climate, you remember that Herschel said this in his letter to me. If I had stated as plainly as he has done the possibility of the introduction or origination of fresh species being a natural, in contradistinction to a miraculous process, I should have raised a host of prejudices against me, which are unfortunately opposed at every step to any philosopher who attempts to address the public on these mysterious subjects."[46]

In a letter to Professor Sedgwick on January 20, 1838, he wrote: "...I have left it rather to be inferred instead of enunciating it even as my opinion, that the place of lost species is filled up (as it was of old) from time to time by new species."[47] "...though *how,* is a point we are as ignorant of as of the man-

[44] *Ibid.,* Vol. I, p. 467.
[45] *Ibid.,* Vol. I, p. 468.
[46] *Ibid.,* Vol. II, p. 5.
[47] *Ibid.,* Vol. II, p. 36.

ner of God's creating the first man."[48] However, "The burden
of proof rests on him who ventures to affirm that Nature has,
at length, stopped short in her operations, and that while the
causes of destruction are in full activity, even where man cannot interfere, she has suspended her powers of repair and
renovation."[49]

Darwin was impatient with Lyell for not plainly stating his
position. On March 6, 1863, he wrote Lyell: "I will first get
out what I hate saying, viz., that I have been greatly disappointed that you have not given judgment and spoken fairly
out what you think about the derivation of species.... I think
the *Parthenon* is right, that you will leave the public in a fog."[50]

Lyell's approach, however, was to get his ideas into the minds
of others without their being conscious at first of the logical
conclusions of his positions.

Lyell Was an Evolutionist

Henshaw Ward maintained that Lyell never became an
evolutionist.[51] There are statements in Lyell's letters, however,
and in the tributes paid to him after his death, that establish
the contrary. Writing to Professor Heer, August 26, 1861, he
expressed a tendency toward some form of evolutionism. "I have
gone much farther than before in favour of a progressive development, and have also endeavoured to show the importance of
botany in geological classification."[52]

His feelings, as well as the feelings of others, held him back
at times. As he expressed it: "My feelings, however, more than
any thought about policy or expediency, prevent me from dog-
matising as to the descent of man from the brutes, which, though
I am prepared to accept it, takes away much of the charm from
my speculations on the past relating to such matters."[53]

---

[48] *Ibid.*, Vol. II, p. 36.
[49] *Ibid.*, Vol. II, p. 37.
[50] *Life and Letters of Charles Darwin*, Vol. II, p. 196.
[51] *Charles Darwin and the Theory of Evolution*, pp. 433-445.
[52] Lyell, *op. cit.*, Vol. II, p. 350.
[53] *Ibid.*, Vol. II, p. 363. March 11, 1863.

His position was made somewhat clearer in a letter to Darwin on January 16, 1865. "I have some notes of it, and hope one day to run over it with you, especially as it was somewhat of a confession of faith as to the 'Origin.' I said I had been forced to give up my old faith without thoroughly seeing my way to a new one. But I think you would have been satisfied with the length I went."[54]

The clearest statement, which these authors have found, is in a letter to Professor Haeckel, November 23, 1868. "But while I taught that as often as certain forms of animals and plants disappeared, for reasons quite intelligible to us, others took their place by virtue of a causation which was beyond our comprehension; it remained for Darwin to accumulate proof that there is no break between the incoming and the outgoing species, that they are the work of evolution, and not of special creation."[55]

Lyell's sister-in-law included, in the *Life, Letters and Journals of Sir Charles Lyell,* some of the tributes that appeared after Lyell's death. Evidently she and others who had been close to him were convinced that Lyell had become an evolutionist. For out of the many tributes she selected such statements as the following. One tribute, in referring to the impact of Darwin's *Origin of Species,* said that: "Whenever Sir Charles Lyell considered that a case had been fairly made out, he was too noble to shut his eyes against the evidence, but freely accepted the new conclusion, even to the overthrow of his previous work."[56] One writer maintained that after advocating ideas of special creation in nine editions of his *Principles of Geology,* that "he put forth a tenth edition, in which the old theory was formally renounced, and the new one taken up."[57] He spoke of "his late conversion to Mr. Darwin's doctrine of Natural Selection."[58]

---

[54] *Ibid.,* Vol. II, p. 384.
[55] *Ibid.,* Vol. II, p. 436.
[56] *Ibid.,* Vol. II, pp. 471-472.
[57] *Ibid.,* Vol. II, p. 473.
[58] *Ibid.,* Vol. II, p. 472.

Dr. Hooker spoke of this change as an act of heroism.[59] And so he who prepared the way for Darwin finally followed Darwin.

---

[59] *Ibid.*, Vol. II, p. 473.

CHAPTER III
CHARLES DARWIN

How was it that Charles Darwin, a believer in Christ and a creationist, came to be an agnostic whose name is inevitably connected today with the hypothesis of evolution? What swayed the man who has swayed so many? Why did Darwin, who was supposed to have prepared himself for the work of a clergyman in the Church of England, become the chief apostle of evolutionism?[1]

Darwin's Plans for the Clergy

Charles Darwin had gone to Cambridge with the intention of becoming a clergyman. In March, 1829, he had doubts about his "call to the ministry," however, though he seems to have overcome them for a while. His son wrote that: "Mr. Herbert's sketch shows how doubts arose in my father's mind as to the possibility of his taking Orders. He writes, 'We had an earnest conversation about going into Holy Orders; and I remember his asking me, with reference to the question put by the Bishop in the ordination service, "Do you trust that you are inwardly moved by the Holy Spirit, etc.," whether I could answer in the affirmative, and on my saying I could not, he said, "Neither can I, and therefore I cannot take Orders." ' This conversation appears to have taken place in 1829, and if so, the doubts here expressed must have been quieted, for in May 1830, he speaks of having some thoughts of reading divinity with Henslow."[2]

---

[1] The best analysis of this question, known to the writers, is found in Robert E. D. Clark, *Darwin: Before and After,* London: The Paternoster Press, 1948. See the chapter on "What Darwin Accomplished."

[2] *Life and Letters of Charles Darwin,* Vol. I, p. 147.

Evidently he did not go very deeply into preparation for the ministry since he never felt that he was capable of writing on religious subjects because he had not thought them out carefully and systematically enough.[3]

The intention of becoming a clergyman seems never to have been definitely renounced, but rather gradually faded away as his interest in natural science grew and as unbelief gradually crept in.[4] Something of the growth of this unbelief is traced for us by Darwin when he said: "During these two years (Oct. 1836 —Jan. 1839) I was led to think much about religion. Whilst on board the Beagle I was quite orthodox, and I remember being heartily laughed at by several of the officers (though themselves orthodox) for quoting the Bible as an unanswerable authority on some point of morality. I suppose it was the novelty of the argument that amused them. But I had gradually come by this time, i.e. 1836 to 1839, to see that the Old Testament was no more to be trusted than the sacred books of the Hindoos. The question then continually rose before my mind and would not be banished,—is it credible that if God were now to make a revelation to the Hindoos, he would permit it to be connected with the belief in Vishnu, Siva, &c., as Christianity is connected with the Old Testament? This appeared to me utterly incredible.

"By further reflecting that the clearest evidence would be requisite to make any sane man believe in the miracles by which Christianity is supported,—and that the more we know of the fixed laws of nature the more incredible do miracles become,[5] —that the men at that time were ignorant and credulous to a degree almost incomprehensible by us,[6]—that the Gospels cannot

---

[3] *Ibid.,* Vol. I, pp. 275-276.

[4] Clark, *op. cit.,* p. 85.

[5] For a discussion of the possibility of miracles see C. S. Lewis, *Miracles: A Preliminary Study,* New York: The Macmillan Co., 1959; W. H. Fitchett, *The Beliefs of Unbelief,* New York: Cassell Co., Ltd., 1911; Frank Ballard, *The Miracles of Unbelief,* 4th Ed., Edinburgh: T. & T. Clark, 38 George St., 1902.

[6] The more credulous a people are, the more difficult it is to convince them of the truth of that which is contrary to their prejudices. The gospel was contrary to their prejudices.

be proved to have been written simultaneously with the events,[7] —that they differ in many important details, far too important, as it seemed to me, to be admitted as the usual inaccuracies of eye-witnesses;[8]—by such reflections as these, which I give not as having the least novelty or value, but as they influenced me, I gradually came to disbelieve in Christianity as a divine revelation. The fact that many false religions have spread over large portions of the earth like wild-fire had some weight with me.[9]

"But I was very unwilling to give up my belief; I feel sure of this, for I can well remember often and often inventing day-dreams of old letters between distinguished Romans, and manuscripts being discovered at Pompeii or elsewhere, which confirmed in the most striking manner all that was written in the Gospels. But I found it more and more difficult, with free scope given to my imagination, to invent evidence which would suffice to convince me. Thus disbelief crept over me at a very slow rate, but was at last complete. The rate was so slow that I felt no distress."[10]

There are several statements, pertinent to the main theme of our book, in the above quotation which need to be considered. *First,* it is noteworthy that Charles Darwin lost his faith in Christianity and the miraculous before he had formulated his hypothesis of evolution. This is not to say that he had no evolutionary ideas before this time; for he did, as shall be shown.[11] But it is still true that he had lost his faith in creation before he set out to discover how life in its varied forms could come about by the working of natural laws. In other words, with

---

[7] See F. F. Bruce, *Are the New Testament Documents Reliable?* London: The Inter-Varsity Fellowship, 1948.

[8] See J. W. McGarvey, *Evidences of Christianity,* Cincinnati: Standard Publishing Co.

[9] See Floyd E. Hamilton, *The Basis of Christian Faith,* New York: Harper and Brothers, 1946, pp. 101-131.

[10] *Life and Letters,* Vol. I, pp. 277-278. Compare John C. Greene, *Darwin and the Modern World View,* New York: The New American Library, 1963, pp. 16-17.

[11] *More Letters of Charles Darwin,* Vol. I, p. 367.

Darwin, as it will be shown to be true of others, evolution came in to fill up the void that was being created by the departure of faith in the God who creates.

Second, although faith in the Old Testament was first shaken, and this was somewhat connected with the so-called "moral difficulties,"[12] evidently something deeper than this was at work, i.e., his faith in the doctrine of uniformity and thus his loss of faith in the miraculous. Of course, if the extreme theory of uniformity be true the entire Bible must be discredited, for it is based on the supernatural manifestations of God.

Third, one of the very significant things about the growth of Darwin's unbelief was the use of his imagination instead of the use of his reason in trying to solve the problems that unbelief had raised in his mind. Instead of searching for the solutions of the problems concerning the New Testament, he imagined solutions. In other words, he did not deal honestly with his difficulties and his faith. Finally he decided to face and to accept his doubts and, since he had not accumulated by the process of imagination any real reasons for faith or any solutions to difficulties, he became an unbeliever. Thus doubt had successfully built up its case against a faith sustained by imagination. Had he been willing to deal with doubt differently, if he had been as diligent in his study of Christian evidences as he was of natural history, his faith might have found a firm basis instead of being destroyed because it was based on fancies.

Fourth, Darwin claimed that he was unwilling to give up his faith. "Yet as Warfield[13] has pointed out, his words really point to the opposite conclusion for he finished by saying: 'I found it

---

[12] See George W. DeHoff, *Alleged Bible Contradictions Explained,* Murfreesboro, Tenn.: DeHoff Publications, 1950; Theodore Engelder, *Scripture Cannot Be Broken,* St. Louis: Concordia Publishing Co.; Floyd E. Hamilton, *The Basis of Christian Faith,* New York: Harper and Brothers, 1946, pp. 263-274; J. W. McGarvey, *Evidences of Christianity,* Cincinnati: Standard Publishing Co., Vol. II; R. A. Torrey, *Difficulties and Alleged Errors and Contradictions in the Bible,* Grand Rapids: Baker Book House, reprinted 1966.

[13] B. B. Warfield, *Studies in Theology,* 1932, pp. 541 ff.

more and more difficult, with free scope given to my imagination, to invent evidence which would convince me.' No remark could reveal more clearly that, while pretending to himself that he wanted to believe, Charles was really determined at all costs *not* to believe and so, in order to rationalize his unbelief, he steadily raised the level of evidence he required before he would be convinced."[14]

Fifth, as Robert E. D. Clark has shown, Darwin's loss of faith bothered him far more than he was willing to admit.[15]

Darwin and Lyell

Darwin recognized the influence of Lyell on him, and he realized that Lyell was viewed "as head of the uniformitarians."[16] "The Science of Geology is enormously indebted to Lyell, more so, as I believe, than to any other man who ever lived."[17]

Darwin was also indebted to Lyell. As he expressed it: "When [I was] starting on the voyage of the Beagle, the sagacious Henslow, who, like all other geologists, believed at that time in successive cataclysms, advised me to get and study the first volume of the 'Principles,' which had then just been published, but on no account to accept the views therein advocated. How differently would any one now speak of the 'Principles!' I am proud to remember that the first place, namely, St. Jage, in the Cape de Verde archipelago, in which I geologized, convinced me of the infinite superiority of Lyell's views over those advocated in any other work known to me."[18] Darwin studied Lyell "attentively; and the book was of the highest service to me in many ways." Not only was his manner of treating geology su-

---

[14] Clark, *op. cit.*, p. 83. See also Clark's discussion of the "economy of evidence" in *Conscious and Unconscious Sin,* London: Williams and Norgate, 1934.

[15] *Ibid.*, p. 84. Compare Nora Barlow, *The Autobiography of Charles Darwin,* London: Collins, 1958, pp. 235-239; Julian Huxley, "The Emergence of Darwinism," Sol Tax, Editor, *The Evolution of Life,* Chicago: University of Chicago Press, 1960, pp. 3-5.

[16] Letter to Lyell, July 30, 1860; *Life and Letters,* Vol. II, p. 121.

[17] *More Letters,* Vol. II, p. 118, footnote.

[18] *Life and Letters,* Vol. I, p. 60.

perior to any that Darwin had read up to that time, but in his
autobiography he stated that it was superior to any he had
"ever afterwards read."[19]

In a letter (Feb. 23, 1875) to Lyell's secretary just after Lyell's
death he paid tribute to Lyell as one who had "revolutionised
Geology." "I never forget that almost everything which I have
done in science I owe to the study of his great works."[20] The
influence of Lyell's doctrine of uniformity, and of other naturalistic ideas in Darwin's mind, was such that he lost faith in
creation and began to search for a naturalistic explanation of the
world of living creatures.[21]

Creation Unscientific

To accept uniformity is to reject creation as unscientific. On
July 20, 1856, he told Asa Gray that: "I *assume* that species
arise like our domestic varieties with *much* extinction; and then
test this hypothesis by comparison with as many general and
pretty well-established propositions as I can find made out,—
in geographical distribution, geological history, affinities, etc.,
etc. And it seems to me that, *supposing* that such hypothesis
were to explain such general propositions, we ought, in accordance with the common way of following all sciences, to admit
it till some better hypothesis be found out. For to my mind to
say that species were created so and so is no scientific explanation, only a reverent way of saying it is so and so."[22]

This being Darwin's attitude it is no wonder that belief in
evolution came as a relief to him. "I believe that Hopkins is
so much opposed because his course of study has never led him
to reflect much on such subjects as geographical distribution,
classification, homologies, etc., so that he does not feel it a
relief to have some kind of explanation."[23] Yet, Darwin left in
the various editions of the *Origin of Species* his statement that

---

19 *Ibid.*, Vol. I, p. 52.
20 *Ibid.*, Vol. II, p. 374.
21 *Ibid.*, Vol. I, pp. 276-285.
22 *Ibid.*, Vol. I, p. 437.
23 *Ibid.*, Vol. II, p. 120. Darwin to Asa Gray, July 22, 1860.

the Creator breathed life into the first form or first few forms of life.

Evolution Accepted Early

There are some who think that Darwin accepted the theory of evolution only after many, many years of studying the subject. This, however, is not the case. As his religious faith ebbed his faith in evolution developed. It came in to fill up the void that was being left by his rejection of creation. And, of course, if anything took the place of the idea of creation it had to be some form of evolution.

During the voyage of the *Beagle* he had been struck by certain things that "could only be explained on the supposition that species gradually become modified; and the subject haunted me."[24] In writing to Otto Zacharias in 1877 he said: "When I was on board the *Beagle* I believed in the permanence of species, but, as far as I can remember, vague doubts occasionally flitted across my mind. On my return home in the autumn of 1836 I immediately began to prepare my journal for publication, and then saw how many facts indicated the common descent of species, so that in July, 1837, I opened a notebook to record any facts which might bear on the question; but I did not become convinced that species were mutable until, I think, two or three years had elapsed."[25]

At another time he wrote: "In October 1838, that is, fifteen months after I had begun my systematic enquiry, I happened to read for amusement 'Malthus on Population,' and being well prepared to appreciate the struggle for existence which everywhere goes on from long-continued observation of the habits of animals and plants, it at once struck me that under these circumstances favourable variations would tend to be preserved, and unfavourable ones to be destroyed. The result of this would be the formation of new species. Here then I had at last got a theory by which to work; but I was so anxious to avoid preju-

---

[24] *Ibid.,* Vol. I, p. 67.
[25] *More Letters,* Vol. I, p. 367; *Life and Letters,* Vol. I, p. 56.

dice, that I determined not for some time to write even the briefest sketch of it. In June 1842 I first allowed myself the satisfaction of writing a very brief abstract of my theory in pencil in 35 pages; and this was enlarged during the summer of 1844 into one of 230 pages, which I had fairly copied out and still possess."[26]

Not Proved

Many people believe that Darwin proved evolution. But Darwin recognized that his hypothesis was beset with difficulties. In writing to T. H. Huxley, December 2, 1860, he said: "I entirely agree with you, that the difficulties on my notions are terrific, yet having seen what all the Reviews have said against me, I have far more confidence in the *general* truth of the doctrine than I formerly had."[27]

Darwin not only recognized that difficulties were in its way, but he also realized that *it had not been proved*. To G. Bentham, on May 22, 1863, he admitted that "In fact, the belief in Natural Selection must at present be grounded entirely on general considerations. (1) On its being a *vera causa*, from the struggle for existence; and the certain geological fact that species do somehow change. (2) From the analogy of change under domestication by man's selection. (3) And chiefly from this view connecting under an intelligible point of view a host of facts. When we descend to details, we can prove that no one species has changed (i.e., we cannot prove that a single species has changed); nor can we prove that the supposed changes are beneficial, which is the ground work of the theory. Nor can we explain why some species have changed and others have not. The latter case seems to me hardly more difficult to understand precisely and in detail than the former case of supposed change."[28] So when one came down to details—and to details science must go—his hypothesis could not be proved.

---

[26] *Life and Letters*, Vol. I, p. 68. Compare Gertrude Himmelfarb, *Darwin and the Darwinian Revolution*, pp. 161 ff.

[27] *Ibid.*, Vol. II, p. 147.

[28] *Ibid.*, Vol. II, p. 210.

In another letter to Bentham, June 19, 1863, he granted that "I, for one, can conscientiously declare that I never feel surprised at any one sticking to the belief of immutability; though I am often not a little surprised at the arguments advanced on this side. I remember too well my endless oscillations of doubt and difficulty."[29]

Darwin's Attitude toward God

In a letter to Sir Charles Lyell on August 21, 1861, Darwin expressed his reluctance to think on whether or not Intelligence had anything to do with the origin of species. "The conclusion which I always come to after thinking of such questions is that they are beyond the human intellect; and the less one thinks on them the better. You may say, Then why trouble me? But I should very much like to know clearly what you think."[30] Being a human being he did think from time to time about the idea of God, and some of these thoughts will be considered in this section.

In all of his intellectual wanderings Darwin never became an atheist. In 1879 he pointed out that "In my most extreme fluctuations I have never been an atheist in the sense of denying the existence of a God. I think that generally (and more and more as I grow older), but not always, that an Agnostic would be the more correct description of my state of mind."[31] To J. D. Hooker on July 12, 1870 he said that "My theology is a simple muddle; I cannot look at the universe as the result of blind chance, yet I can see no evidence of beneficent design, or indeed of design of any kind, in the details."[32] Yet he never claimed to be an atheist and his nephew, Francis Darwin, thought that those who classified him as such abused and misrepresented him.[33]

Darwin was not only unwilling to be classified as an atheist,

---

[29] *Ibid.*, Vol. II, pp. 210-211.
[30] *More Letters of Charles Darwin*, Vol. I, p. 194.
[31] *Life and Letters*, Vol. I, p. 274.
[32] *More Letters*, Vol. I, p. 321.
[33] *Ibid.*, Vol. I, p. 258 footnote; *Life and Letters*, Vol. I, p. 286.

but he also recognized that faith in God was reasonable. In his autobiography, written in 1876, he put it this way: "Another source of conviction in the existence of God, connected with the reason, and not with the feeling, impresses me as having much more weight. This follows from the extreme difficulty or rather impossibility of conceiving this immense and wonderful universe, including man with his capacity of looking far backwards and far into futurity, as the result of blind chance or necessity. When thus reflecting I feel compelled to look to a First Cause having an intelligent mind in some degree analogous to that of man; and I deserve to be called a Theist. This conclusion was strong in my mind about the time, as far as I can remember, when I wrote the 'Origin of Species;' and it is since that time that it has very gradually, with many fluctuations, become weaker. But then arises the doubt, can the mind of man, which has, as I fully believe, been developed from a mind as low as that possessed by the lowest animals, be trusted when it draws such grand conclusions?"[34]

Again he wrote: "Nevertheless you have expressed my inward conviction, though far more vividly and clearly than I could have done, that the Universe is not the result of chance. But then with me the horrid doubt always arises whether the convictions of man's mind, which has been developed from the mind of the lower animals, are of any value or at all trustworthy. Would any one trust in the convictions of a monkey's mind, if there are any convictions in such a mind?"[35]

Reason led Darwin to God, so Darwin killed reason. He trusted his mind when reasoning about evolution, but not about God! What a warning from the author to the reader, this discrediting of reason would have made as a preface to the *Origin of Species* and *The Descent of Man!*

Darwin again returned to this subject and said: "On the other hand, if we consider the whole universe, the mind refuses to look at it as the outcome of chance—that is, without design or

---

[34] *Life and Letters,* Vol. I, p. 282.
[35] *Ibid.,* Vol. I, p. 285.

purpose. The whole question seems to me insoluble, for I cannot put much or any faith in the so-called intuitions of the human mind, which have been developed, as I cannot doubt, from such a mind as animals possess; and what would their convictions or intuitions be worth?"[36]

If our minds are as he describes them, and our convictions untrustworthy, why does he say that "I cannot doubt" the evolution of the mind of man from such as the animals possess? How could he trust his mind when it thought on the theory of evolution? As Arnold Lunn put it: "A clear thinker would never have been guilty of such inconsistent reasoning. If Darwin was not prepared to trust his mind when it drew the 'grand conclusion' that God existed, why was he prepared to trust it when it drew the depressing conclusion that a mind of such bestial origin could not be trusted to draw any conclusion at all?

"Darwin's mind at different periods of his life led him to two firm convictions: (a) that God exists, and (b) that man is descended from the lower animals.

"If, as the result of (b) he lost confidence in his own mental processes, he might well have rejected both beliefs; but to retain the latter belief, which was the source of his Scepticism, and to reject the former was illogical. It was, indeed, absurd to state on the same page that he 'fully believed' in the bestial origin of his own mind, and that this same bestial origin did not entitle him 'fully to believe' in anything."[37]

Why should Darwin have trusted the human mind when it drew the sweeping conclusion that the world had had its origin and development by a process of evolution, which he admitted had not been proved?[38] In fact, why should he have accepted any conclusions of human reason? Why did he not use the same argument against positions in addition to belief in God? How trustworthy was his argument against reason?

---

[36] *More Letters,* Vol. I, p. 395. August 28, 1881.
[37] Arnold Lunn, *The Revolt against Reason,* N.Y.: Sheed and Ward, 1951, p. 151.
[38] *Life and Letters,* Vol. II, pp. 210-211.

Darwin, it is clear, felt compelled to admit belief in God because of the utter impossibility of the atheistic view. Reason demanded the theistic position. And yet, he refused to accept the position. Why? Because it was unreasonable? No, for its reasonableness was acknowledged by him. It must have been because of some violent prejudice since rather than to draw the reasonable conclusion, he discredited human reason itself. What an unreasonable reaction! When a man does something like this one can be assured that it is something besides reason which is leading him to draw his conclusions.

He should have first settled the question as to whether or not monkeys have convictions. If they do, one could then endeavor to determine how trustworthy they are. If they do not, one does not need to answer the question for the question has no meaning. Obviously if they have no convictions, one could not speak of trusting or distrusting what they did not have.

Furthermore, monkeys do get some things right, although how they do it is another question. We would be ready to trust the convictions of a monkey concerning God if he was capable of having and expressing such convictions, and if he upheld them with reasonable arguments. But since the monkey does not do this, and man does, it is best to deal with man's reasonable conclusions instead of dragging a monkey into a discussion where he does not belong, and into which he has not given any indication that he wants to be invited.

In other connections Darwin indicated that man should feel proud in having risen so high from such a humble beginning,[39] but when questions concerning God arose he ditched the "pride" and humbled himself to the monkey's level of reasoning,[40] and then discredited the monkey. Darwin's determination not to believe cost him his mind!

Robert E. D. Clark's comments concerning Darwin's monkey illustration are penetrating. "As his correspondence shows, such

---

[39] *Descent of Man,* Vol. I, p. 221.
[40] Compare Robert E. D. Clark, *Darwin: Before and After,* pp. 77-80.

thoughts arose in his mind whenever he was faced with theological issues, yet they never worried him in other connections. He never doubted natural selection on the ground that, if an animal at the Zoo had become an orthodox Darwinian, no one would have taken any notice.

"By such reasoning, Darwin allowed his evolutionary views to destroy all serious thinking about ultimate issues. Yet, once again he never seems to have realized for an instant that these all-pervading doubts were in no way necessarily connected with man's ancestry. We do not take the conclusions of a new-born baby or of a child very seriously, but no one thinks of arguing that philosophers are not to be trusted because they are only grown-up babies. Yet it was precisely this consideration, applied to monkeys instead of babies, that bothered Charles Darwin.

"This new excuse for all-pervading doubt which Darwin suggested to his contemporaries had a considerable vogue. Indeed, it still seems to be imagined in some quarters that it forms part and parcel of evolutionary thinking. Of late years there has been a reaction to the doctrine that the proof of the pudding is not in the eating but in the cook's pedigree, and the excuse has become more disguised in character. Yet, when modern rationalist writers inform us that the human mind is unfit (not sufficiently evolved?) to discuss transcendental problems, they are really harking back to the old difficulty which Charles Darwin raised. Like Charles they fail to observe that the same difficulty might be raised against thinking at any level."[41]

The question of God's existence was still on Darwin's mind during the last year of his life. "The Duke of Argyll[42] has recorded a few words on this subject, spoken by my father in the last year of his life. '... in the course of that conversation I said to Mr. Darwin, with reference to some of his own remarkable works on the "Fertilization of Orchids," and upon "The Earthworms," and various other observations he made of the wonderful contrivances for certain purposes in nature—I said

---

[41] *Ibid.,* pp. 80-81.
[42] "Good Words," April, 1885, p. 244.

it was impossible to look at these without seeing that they were the effect and the expression of mind. I shall never forget Mr. Darwin's answer. He looked at me very hard and said, "Well, that often comes over me with overwhelming force; but at other times," and he shook his head vaguely, adding, "it seems to go away." ' "[43]

Small wonder that it went away, and grew weaker. What else could happen when he so suppressed his convictions and so discredited human reason itself in order to escape an extremely reasonable conclusion? What happened to him in this case may, in some measure, be illustrated by what happened when Darwin neglected poetry and music. Darwin was at one time a great lover of music and of poetry. And yet neglect of these things, not opposition to them, finally led to loss of appreciation for them. He put it this way: "I have said that in one respect my mind has changed during the last twenty or thirty years. Up to the age of thirty, or beyond it, poetry of many kinds, such as the works of Milton, Gray, Byron, Wordsworth, Coleridge, and Shelley, gave me great pleasure, and even as a schoolboy I took intense delight in Shakespeare, especially in the historical plays. I have also said that formerly pictures gave me considerable, and music very great delight. But now for many years I cannot endure to read a line of poetry: I have tried lately to read Shakespeare, and found it so intolerably dull that it nauseated me. I have also almost lost my taste for pictures or music. Music generally sets me thinking too energetically on what I have been at work on, instead of giving me pleasure. I retain some taste for fine scenery, but it does not cause me the exquisite delight which it formerly did. On the other hand, novels which are works of the imagination, though not of a very high order, have been for years a wonderful relief and pleasure to me, and I often bless all novelists. A surprising number have been read aloud to me, and I like all if moderately good, and if they do not end unhappily—against which a law ought to be passed. A novel, according to my taste, does not come into the first

---

[43] *Life and Letters,* Vol. I, p. 285, footnote.

class unless it contains some person whom one can thoroughly love, and if a pretty woman all the better.

"This curious and lamentable loss of the higher aesthetic tastes is all the odder, as books on history, biographies, and travels (independently of any scientific facts which they may contain), and essays on all sorts of subjects interest me as much as ever they did. My mind seems to have become a kind of machine for grinding general laws out of large collections of facts, but why this should have caused the atrophy of that part of the brain alone, on which the higher tastes depend, I cannot conceive. A man with a mind more highly organised or better constituted than mine, would not, I suppose, have thus suffered; and if I had to live my life again, I would have made a rule to read some poetry and listen to some music at least once every week; for perhaps the parts of my brain now atrophied would thus have been kept active through use. The loss of these tastes is a loss of happiness, and may possibly be injurious to the intellect, and more probably to the moral character, by enfeebling the emotional part of our nature."[44]

After the publication of the *Life and Letters of Charles Darwin* by his son, the *Atlantic Monthly,* in reviewing the book, draws up an estimate of the religious life of the great transmutationist:

"The blank page in this charming biography is the page of spiritual life. There is nothing written there. The entire absence of an element which enters commonly into all men's lives in some degree is a circumstance as significant as it is astonishing .... The spiritual element in life is not remote, but it is not a matter of sensation, and Darwin lived as if there were no such thing; it belongs to the region of emotion and imagination, and those perceptions which deal with the nature of man in its contrast with the material world. Poetry, art, music—Darwin's insensibility to the higher life—for so men agree to call it—was partly if not wholly, induced by his absorption in scientific pursuits in the spirit of materialism.... Great as Darwin was

---

[44] *Ibid.,* Vol. I, pp. 81-82.

as a thinker, and winning as he remains as a man, those elements in which he was deficient are the noblest part of our nature.

"On finishing the story of his life, the reflection rises involuntarily in the mind that this man, in Wordsworth's line, 'hath faculties that he has never used.' "[45]

Darwin had not only neglected, but had actually opposed and suppressed his convictions concerning God. Small wonder that these convictions wavered and often became weaker. And yet, throughout life it may be that this suppression of his religious convictions led to a feeling of guilt and even to frequent illness. But this is another story which has been discussed by Robert E. D. Clark.[46]

Evolution Compatible with Faith in God

That Darwin believed that faith in the theory of evolution was compatible with faith in God is evident from the fact that he was not an atheist and admitted that faith in God was reasonable. In writing to a German student in 1879 (one of his family did the writing for him) he said that "He considers that the theory of Evolution is quite compatible with the belief in a God; but that you must remember that different persons have different definitions of what they mean by God."[47]

God may have created the first form or forms of life, Darwin suggested on the last page of *Origin of Species*, but he did not think that God had intervened in the various stages of evolution. This, he said, would have made the theory of natural selection valueless.[48] Although he left the statement in the *Origin of Species*, Darwin's faith in evolution was simply an extension of his faith in uniformity and, logically enough, this same faith led him to believe in spontaneous generation of a living creature

---

[45] "Darwin's Life," *Atlantic Monthly*, April, 1888, p. 560. Quoted by Mary Frederick, *Religion and Evolution Since 1859*, University of Notre Dame, 1934, p. 24.

[46] *Darwin: Before and After*, pp. 84-90.

[47] *Life and Letters*, Vol. I, p. 277.

[48] *Ibid.*, Vol. I, p. 528; Vol. II, p. 6.

from inorganic matter. Not evidence, but the doctrine of uniformity, led him to this conclusion. Thus he wrote to D. Mackintosh, February 28, 1882, that "Though no evidence worth anything has as yet, in my opinion, been advanced in favour of a living being, being developed from inorganic matter, yet I cannot avoid believing the possibility of this will be proved some day in accordance with the law of continuity."[49]

Faith in God Weakened

Darwin recognized that his hypothesis of evolution flowed away from, rather than towards, God. On August 8, 1860, in a letter to Huxley he spoke of him as "My good and kind agent for the propagation of the Gospel—i.e. the devil's gospel."[50] As Sedgwick saw, Darwinism helped to further brutalize mankind through providing "scientific sanction" for blood-thirsty and selfish desires. On December 24, 1859, he told Darwin: "If Darwinism did what he thought it could do, humanity, in my mind, would suffer a damage that might brutalize it, and sink the human race into a lower grade of degradation than any into which it has fallen since its written records tell us of its history."[51]

Paley and Darwin

Although some writers, such as Asa Gray[52] in America, believed that Darwinism did not undermine Paley's argument from design for the existence of God, yet Darwin thought that it did. In fact, Clark[53] is persuaded that one of the reasons that Darwin was so deeply interested in natural selection was that it enabled him to avoid the force of Paley's *Natural Theology*. Darwin

---

[49] *More Letters,* Vol. II, p. 171. For a discussion of this question see James D. Bales, *Communism: Its Faith and Fallacies,* Grand Rapids 6, Michigan: Baker Book House, 1962.

[50] *Life and Letters,* Vol. II, p. 124.

[51] *Ibid.,* Vol. II, p. 44. For the way in which it was used the interested reader is referred to R. Hofstadter's *Social Darwinism.*

[52] Clark, *op. cit.,* p. 105.

[53] *Ibid.,* p. 86.

was, in effect, running from God as is evident from his treatment of reason when it led him to God.

Darwin had once been an admirer of Paley. "I do not think I hardly ever admired a book more than Paley's 'Natural Theology.' I could almost formerly have said it by heart."[54] Yet, he was persuaded, Paley fell with the advent of natural selection. "The old argument from design in Nature, as given by Paley, which formerly seemed to me so conclusive, fails, now that the law of natural selection has been discovered.... There seems to be no more design in the variability of organic beings, and in the action of natural selection, than in the course which the wind blows."[55]

In writing to Asa Gray, November 26, 1860, Darwin expressed regret that he could not go as far as Gray with reference to design. "But I grieve to say that I cannot honestly go as far as you do about Design. I am conscious that I am in an utterly hopeless muddle. I cannot think that the world, as we see it, is the result of chance; and yet I cannot look at each separate thing as the result of Design.... Again, I say I am, and shall ever remain, in a hopeless muddle."[56]

Evidently Gray wanted to know just what it would take to lead Darwin to believe in design, and Darwin's answer indicated that it would be well nigh impossible for him to be convinced. Darwin adopted an either-or attitude and thus because he could not conceive of his nose being designed he rejected all design. He did this even after he had admitted that it was impossible for him to believe that all was but the work of chance. Under the date of September 17, 1861, he wrote Asa Gray that: "Your question what would convince me of Design is a poser. If I saw an angel come down to teach us good, and I was convinced from others seeing him that I was not mad, I should believe in design. If I could be convinced thoroughly

---

[54] *Life and Letters,* Vol. II, p. 15. Letter Nov. 15, 1859, to John Lubbock.

[55] *Ibid.,* Vol. I, pp. 278-279.

[56] *Ibid.,* Vol. II, p. 146.

that life and mind was in an unknown way a function of other imponderable force, I should be convinced. If man was made of brass or iron and no way connected with any other organism which had ever lived, I should perhaps be convinced. But this is childish writing.

"I have lately been corresponding with Lyell, who, I think, adopts your idea of the stream of variation having been led or designed. I have asked him (and he says he will hereafter reflect and answer me) whether he believes that the shape of my nose was designed. If he does I have nothing more to say. If not, seeing what Fanciers have done by selecting individual differences in the nasal bones of pigeons, I must think that it is illogical to suppose that the variations, which natural selection preserves for the good of any being have been designed. But I know that I am in the same sort of muddle (as I have said before) as all the world seems to be in with respect to free will, yet with everything supposed to have been foreseen or preordained."[57]

In dealing with the argument from design Darwin adopted something of the procedure that A. R. Wallace had used in another connection, i.e., he stated the position in such an extreme form that it only had to be stated to be rejected. Either all was designed in the most minute events and details or nothing was designed. Darwin adopted the either-or fallacy and concluded that there was no other alternative. Thus he wrote to Charles Lyell, August 21, 1861, "Will you honestly tell me (and I should be really much obliged) whether you believe that the shape of my nose (eheu!) was ordained and 'guided by an intelligent cause?' "[58] "As for each variation that has ever occurred having been preordained for a special end, I can no more believe in it than that the spot on which each drop of rain falls has been specially ordained."[59]

---

[57] *Ibid.*, Vol. II, pp. 169-170. See also *More Letters*, Vol. I, pp. 193-194.

[58] Compare *Life and Letters*, Vol. II, p. 170.

[59] *Ibid.*, Vol. I, p. 321. To J. D. Hooker, July 12, 1870.

To Dr. Gray (July, 1860) he wrote: "One word more on 'designed laws' and 'undesigned results.' I see a bird which I want for food, take my gun and kill it, I do this *designedly*. An innocent and good man stands under a tree and is killed by a flash of lightning. Do you believe (and I really should like to hear) that God *designedly* killed this man? Many or most persons do believe this; I can't and don't. If you believe so, do you believe that when a swallow snaps up a gnat that God designed that that particular swallow should snap up that particular gnat at that particular instant? I believe that the man and the gnat are in the same predicament. If the death of neither man nor gnat are designed, I see no reason to believe that their *first* birth or production should be necessarily designed."[60]

We certainly understand how the present laws of nature can kill a man, but the present laws of nature give no evidence that they can create a man. An accident can destroy a car but it cannot create a car. A car can be the result of design and its destruction can be an accident.

In this connection we shall present one of Darwin's letters and an extensive comment on it by Dr. Clark. It was to J. D. Hooker in 1862 and deals with a problem that he had raised.

"But the part of your letter which fairly pitched me head over heels with astonishment, is that where you state that every single difference which we see might have occurred without any selection. I do and have always fully agreed; but you have got right round the subject, and viewed it from an entirely opposite and new side, and when you took me there I was astounded. When I say I agree, I must make the proviso, that under your view, as now, each form long remains adapted to certain fixed conditions, and that the conditions of life are in the long run changeable; and second, which is more important, that each individual form is a self-fertilising hermaphrodite, so that each hair-breadth variation is not lost by intercrossing. Your manner of putting the case would be even more striking

---

[60] *Ibid.*, Vol. I, p. 284.

than it is if the mind could grapple with such numbers—it is grappling with eternity—think of each of a thousand seeds bringing forth its plant, and then each a thousand. A globe stretching to the furthest fixed star would very soon be covered. I cannot even grapple with the idea, even with races of dogs, cattle, pigeons, or fowls, and here all admit and see the accurate strictness of your illustration.

"Such men as you and Lyell thinking that I make too much of a Deus of Natural Selection is a conclusive argument against me. Yet I hardly know how I could have put in, in all parts of my book, stronger sentences. The title, as you once pointed out, might have been better. No one ever objects to agriculturists using the strongest language about their selection, yet every breeder knows that he does not produce the modification which he selects. My enormous difficulty for years was to understand adaptation, and this made me, I cannot but think, rightly, insist so much on Natural Selection. God forgive me for writing at such length; but you cannot tell how much your letter has interested me, and how important it is for me with my present book in hand to try and get clear ideas."

Now for Dr. Clark's comment: "In 1862 J. D. Hooker, the botanist, challenged Darwin's claim that natural selection was in any sense a creative agency. Hooker's letter is lost, but Darwin's replies [indicate] it is clear that he said something like this: 'Your theory of evolution by natural selection implies that if every organism had survived and produced off-spring, then every kind of plant and animal that exists and has ever existed would have been produced without any natural selection at all (as well, of course, as myriads of others). In other words, all the characters present in all organisms were the necessary consequences of the earliest and most primitive organism.'

"Darwin had never thought of this before. For a few anxious days, he realized that Paley could not be disposed of as easily as he had imagined. He 'was fairly pitched head over heels with astonishment.' Yet to Hooker's claim that 'every single difference which we see might have occurred without any selection,' he could only go on to say: 'I do and have always fully agreed.'

"Thus, accepting Hooker's argument, Darwin was forced towards the view that the earliest organisms, though apparently so small and simple, were really so gigantically complex that they contained the potentiality of producing all the other organisms that would ever exist on earth. It followed, therefore, that if true the theory of evolution would not abolish Paley's argument from design, but would reinforce it a hundredfold. No wonder Darwin was disturbed. He had sought to escape from God: now he found his old Enemy waiting for him in a new hiding place. His confusion can scarcely be exaggerated. In letter after letter he made the lamest excuses for his inability to think clearly. Intellectually, he said, he was in 'thick mud.' Eventually he tried to avoid the dilemma with a laugh. If everything was designed, then the shape of his nose must have been designed also. (Darwin felt rather sore about the shape of his nose.) So he challenged all and sundry to say whether his nasal profile was designed by the Almighty. And *if* the world was designed, then it followed that God had preordained that a particular bird should swallow a particular gnat at a particular moment in a particular place. All of which was too much for Mr. Darwin to believe. So clearly the world was not designed.

"Perhaps there was never a better instance of a man throwing away the baby with the bath water. Darwin was determined to escape from design and a personal God at all costs. He did so by deciding that either *every* trivial detail in nature must be designed *or else* that there was no design at all. Since the former possibility did not ring true, he refused to discuss the subject seriously any more. As Raven has so well remarked, 'His letters exhibit a resolution not to follow his thoughts to their logical conclusion.' "[61] This is as unreasonable as saying that all we do is planned or nothing is planned.

This chapter has presented, mainly in Darwin's own words, something of his basic attitudes which attracted him to the

---

[61] *Darwin: Before and After,* pp. 88-89.

hypothesis of evolution and which led him finally to accept it. With him, as with others, the doctrine of uniformity led him to deny the miraculous in the Bible—and thus creation as set forth therein—and to accept some naturalistic explanation of the origin of all the various forms of life.

Now let us consider Herbert Spencer.

CHAPTER IV

HERBERT SPENCER

Herbert Spencer was an evolutionist long before Huxley was converted. In speaking of this Huxley observed that: "Outside these ranks, the only person known to me whose knowledge and capacity compelled respect, and who was, at the same time, a thorough-going evolutionist, was Mr. Herbert Spencer, whose acquaintance I made, I think, in 1852, and then entered into the bonds of a friendship which, I am happy to think, has known no interruption. Many and prolonged were the battles we fought on this topic. But even my friend's rare dialectic skill and copiousness of apt illustration could not drive me from my agnostic position."[1]

The enormous influence that Spencer exercised on his generation is indicated, among other things, by the tremendous number of books which were circulated by his publisher in the United States alone. "It may be added here that from the beginning until December 31st, 1903, the Messrs. Appleton have sold 368,755 volumes of Mr. Spencer's writings, but these figures, of course, take no account of the sale of unauthorized editions during the years previous to the adoption of International Copyright."[2] Publishers cannot continue to publish an author's books in such quantities unless there is a big demand for them. And when people buy books it is quite likely that they read at least some of them. This gives one an idea of how tremendous was

[1] *Life and Letters of T. H. Huxley,* Vol. I, p. 242.

[2] *An Autobiography,* by Herbert Spencer. New York: D. Appleton & Co., 1904, Vol. II, p. 113, footnote.

the influence of Spencer in the latter part of the nineteenth century.

Spencer Early Adopted a Naturalistic View

David Duncan, from whom Spencer obtained the promise that he would write the story of his life,[3] informs us that in his childhood Spencer underwent experiences that encouraged him to adopt the naturalistic view of things.

"For the years from seven to thirteen one is dependent mainly on the *Autobiography* and on memoranda by his father. Written late in life, the father's reminiscences could not fail to reflect in some measure the consciousness of the eminence the son had attained to, and Spencer's own recollections could not but be coloured by interpretations derived from subsequent experience. Little progress was made in routine school lessons, yet he acquired an unusual amount of miscellaneous information. When barely eleven he attended Dr. Spurzheim's lectures on Phrenology. Before thirteen he assisted his father in preparing experiments in physics and chemistry for teaching purposes. With insect and plant life he had an acquaintance far in advance of other boys, and was skilled in sketching from Nature. Works of fiction were perused with zest. Left much to himself, the tendency to dwell with his own thoughts was strengthened. On the intellectual side one of the chief results of his father's training was the habit it fostered of ever seeking an explanation of phenomena, instead of relying on authority—of regarding everything as naturally caused, and not as the result of supernatural agency. On the moral side its weakest feature was the encouragement it gave to the inherent tendency of a headstrong boy to set authority at defiance."[4]

That Duncan has correctly diagnosed the origin and development of this naturalistic view is abundantly substantiated, and

---

[3] David Duncan, *Life and Letters of Herbert Spencer*. New York: D. Appleton and Co., 1908, Vol. I, p. vii.

[4] *Ibid.*, Vol. I, pp. 13-14.

further elaborated, in a statement from Spencer's own pen. "The nature thus displayed was rather strengthened than otherwise by my father's habit of speculating about causes, and appealing to my judgment with the view of exercising my powers of thinking. By occasional questions of this kind he strengthened that self-asserting nature of which he had at other times reason to complain, but he did not apparently perceive this. Meanwhile he cultivated a consciousness of Cause—made the thought of Cause a familiar one. The discovery of cause is through analysis—the pulling to pieces of phenomena for the purpose of ascertaining what are the essential connexions among them. Hence one who is in the habit of seeking causes is in the habit of analysing. I have up to this time regarded my father as more synthetic than analytic: being led to do so by his perpetual occupation with synthetic geometry. But now, on reconsidering the facts, I see that he was in large measure analytic. He was a great adept of making solutions of puzzles, verbal or physical; and this evidently implies analysis. Moreover, that analysis of articulations implied by his system of shorthand, exhibited the faculty.

"No doubt this habit of mind, inherited from him and fostered by him, flourished the more in the absence of the ordinary appeals to supernatural causes. Though my father retained the leading religious convictions, yet he never appeared to regard any occurrences as other than natural. It should also be remarked that dogmatic teaching played small part in my education. Linguistic culture is based on authority, and as I rebelled against it, the acceptance of things simply on authority was not habitual. On the other hand, the study of Mathematics (conspicuously Geometry and Mechanics), with which my youth was mainly occupied, appeals, at each step in a demonstration, to private judgment, and in a sense recognises the right of private judgment. Many times, too, I assisted in experiments with the air-pump and the electrical machine; so that ideas of physical causation were repeatedly impressed on me. Moreover such small knowledge of natural history as I gained by rearing insects, tended to familiarise me with natural genesis."[5]

The conviction that everything must be explained naturally became so strong with Spencer that his father claimed that his son regarded natural laws in the same way others regarded revealed religion. This statement of his father's was quoted by Spencer in order to show that his father's views were not Spencer's. "From what I see of my son's mind, it appears to me that the laws of nature are to him what revealed religion is to us, and that any wilful infraction of those laws is to him as much a sin as to us is disbelief in what is revealed. And so long as he makes a holy use of his present knowledge, it is my privilege to believe that he will be led into all truth."[6] In commenting on this Spencer observed that although his father did have "certain naturalistic proclivities of thought" he did not entertain a definite naturalistic system of philosophy.

Spencer's bias against the supernatural was so strong that he could not think about the possibility of a supernatural manifestation. In speaking of certain experiences he wrote: "And simple induction would I think almost have led me to believe in supernatural agency were it not that with me the conviction of natural causation is so strong that it is impossible to think away from it."[7]

"If I find myself obliged to hold that there are supernatural manifestations and a supernatural interference with the order of things, then my personal experience would force me to the conclusion that the power underlying things is diabolical."[8]

It was this same attitude of the impossibility of the supernatural that led Spencer to accept evolution. Like Darwin and Huxley, Spencer had rejected the idea of creation and was looking around for something to take its place when he fell in with

---

[5] *Ibid.*, Vol. II, pp. 305-306.

[6] Spencer, *An Autobiography*, Vol. I, p. 655.

[7] To the Countess of Pembroke, January 19, 1896, in Duncan, *Life and Letters of Herbert Spencer*, Vol. II, p. 85.

[8] *Ibid.*, Vol. II, p. 85. Letter of January 21, 1896.

the hypothesis of evolution. In speaking of his paper, "The Development Hypothesis," Spencer commented: "It shows that in 1852 the belief in organic evolution had taken deep root, and had drawn to itself a large amount of evidence—evidence not derived from numerous special instances but derived from the general aspects of organic nature, and from the necessity of accepting the hypothesis of Evolution when the hypothesis of Special Creation has been rejected. The Special Creation belief had dropped out of my mind many years before, and I could not remain in a suspended state: acceptance of the only conceivable alternative was peremptory."[9] "From this time onwards the evolutionary interpretation of things in general became habitual, and manifested itself in curious ways."[10]

Spencer's Admission of His Strong Bias

After admitting that the earlier part of the evidence for evolution was missing and that the remainder was fragmentary and obscure, after saying that it was but a hypothesis and that it probably would never be anything more, Spencer still accepted it. And in the same paragraph his real reason for accepting it was shown to be an anti-supernatural bias that made creation intolerable to him and left him without any position unless he accepted evolution. He accepted evolution not because of evidence, but because it was the only thing that he could accept after rejecting creation. The following statement was written in 1855, four years before Darwin published the *Origin of Species*: "Save for those who still adhere to the Hebrew myth, or to the doctrine of special creations derived from it, there is no alternative but this hypothesis or no hypothesis. The neutral state of having no hypothesis, can be completely preserved only so long as the conflicting evidences appear exactly balanced: such a state is one of unstable equilibrium, which can hardly be permanent. For myself, finding that there is no positive evidence of special creations, and that there is *some* positive evidence of

---

[9] *Ibid.*, Vol. II, p. 319.
[10] *Ibid.*, Vol. II, p. 319.

evolution—alike in the history of the human race, in the modifications undergone by all organisms under changed conditions,
in the development of every living creature—I adopt the hypothesis until better instructed: and I see the more reason for
doing this, in the facts, that it appears to be the unavoidable
conclusion pointed to by the foregoing investigations, and that
it furnishes a solution of the controversy between the disciples
of Locke and those of Kant."[11]

This anti-supernatural bias was built up before Spencer accepted a theory of evolution. In fact, this bias prepared the soil
for the seed of evolution. In speaking of the period from 1838-
40, when he was between eighteen and twenty years of age, he
said: "My father's letters written during this period from time to
time called my¯attention to religious questions and appealed to
religious feelings—seeking for some response. So far as I can remember they met with none, simply from inability to say anything which would be satisfactory to him, without being insincere.

"How had this state of mind unlike that general throughout
our family, arisen? There were, probably, several causes. In childhood the learning of hymns, always, in common with other rote-
learning, disagreeable to me, did not tend to beget any sympathy with the ideas they contained; and the domestic religious
observances on Sunday evenings, added to those of the day,
instead of tending to foster the feeling usually looked for, did
the reverse. As already indicated in Part I, my father had,
partly no doubt by nature and partly as a result of experience, a
repugnance to priestly rule and priestly ceremonies. This repugnance I sympathized with: my nature being, indeed, still more
than his perhaps, averse to ecclesiasticism. Most likely the aver-
sion conspired with other causes to alienate me from ordinary
forms of religious worship.

"Memory does not tell me the extent of my divergence from
current beliefs. There had not taken place any pronounced

---

[11] Herbert Spencer, *The Principles of Psychology*. New York: D.
Appleton and Company, 1897, Vol. I, p. 466, footnote.

rejection of them, but they were slowly losing their hold. Their hold had, indeed, never been very decided: 'the creed of Christendom' being evidently alien to my nature, both emotional and intellectual. To many, and apparently to most, religious worship yields a species of pleasure. To me it never did so; unless, indeed, I count as such the emotion produced by sacred music.... But the expressions of adoration of a personal being, the utterance of laudations, and the humble professions of obedience, never found in me any echoes."[12]

Another factor also was at work. "There was, I believe, a further reason—one more special to myself than are those which usually operate. An anecdote contained in the account of my early life in Hinton, shows how deeply rooted was the consciousness of physical causation. It seems as though I knew by intuition the necessity of equivalence between cause and effect—perceived, without teaching, the impossibility of an effect without a cause appropriate to it, and the certainty that an effect, relevant in kind and in quantity to a cause, must in every case be produced. The acquisition of scientific knowledge, especially physical, had co-operated with the natural tendency thus shown; and had practically excluded the ordinary idea of the supernatural. A breach in the course of causation had come to be, if not an impossible thought, yet a thought never entertained. Necessarily, therefore, the current creed became more and more alien to the set of convictions gradually formed in me, and slowly dropped away unawares. When the change took place it is impossible to say, for it was a change having no marked stages. All which now seems clear is that it had been unobtrusively going on during my stay at Worcester."[13] Thus during the period between eighteen and twenty years of age Spencer was building up a frame of mind that made ridiculous the idea of creation, and which necessitated some sort of hypothesis of evolution to explain the origin and development of life without

---

[12] Spencer, *An Autobiography,* Vol. I, pp. 170-171.
[13] *Ibid.,* Vol. I, pp. 172-173.

a break in the chain of physical causation—thus without super-natural intervention.

Between the ages twenty and twenty-one he read Lyell's *Principles of Geology*. "I had during previous years been cognizant of the hypothesis that the human race has been developed from some lower race; though what degree of acceptance it had from me memory does not say. But my reading of Lyell, one of whose chapters was devoted to a refutation of Lamarck's views concerning the origin of species, had the effect of giving me a decided leaning to them. Why Lyell's arguments produced the opposite effect of that intended, I cannot say. Probably it was that the discussion presented, more clearly than had been done previously, the conception of the natural genesis of organic forms. The question whether it was or was not true was more distinctly raised. My inclination to accept it as true, in spite of Lyell's adverse criticism, was, doubtless, chiefly due to its harmony with that general idea of the order of Nature towards which I had, throughout life, been growing. Supernaturalism, in whatever form, had never commended itself. From boyhood there was in me a need to see, in a more or less distinct way, how phenomena, no matter of what kind, are to be naturally explained. Hence, when my attention was drawn to the question whether organic forms have been specially created, or whether they have arisen by progressive modifications, physically caused and inherited, I adopted the last supposition; inadequate as was the evidence, and great as were the difficulties in the way. [Lamarckianism is generally discredited today, J.D.B.] Its congruity with the course of procedure throughout things at large, gave it an irresistible attraction; and my belief in it never afterwards wavered, much as I was, in after years, ridiculed for entertaining it.

"The incident illustrates the general truth that the acceptance of this or that particular belief, is in part a question of the type of mind. There are some minds to which the marvellous and unaccountable strongly appeal, and which even resent any attempt to bring the genesis of them within comprehension. There are other minds which, partly by nature and partly by culture, have been led to dislike a quiescent acceptance of the unintelligible; and which push their explorations until causation has been carried to its confines. To this last order of minds mine, from the beginning, belonged."[14] Spencer's frame of mind was such that it led him to disregard certain evidence since an examination of the principle of causation carries one to the place where, as Spencer admitted, one has to postulate a First Cause.[15]

It has been established, with Spencer being the witness, that his rejection of creation was due to a bias concerning physical causation, which in turn prepared his mind for, in fact forced him to, a hypothesis of evolution. And Spencer, we must not forget, had a great deal of influence in popularizing the theory; thus his bias indirectly led many to believe in the hypothesis.

Spencer's bias against the supernatural is only one illustration —although possibly the most important one—of his prejudices that sorely hampered his thinking. There are others. Mr. Robertson, a freethinker who was thus in sympathy with Spencer's anti-supernaturalism, recognized this characteristic of Spencer. In speaking of E. B. Tylor's doctrine, he wrote: "Tylor's doctrine, then, can fairly be described as twofold, and not unified; and Spencer, who so often confessed that in general he never read his predecessors,[16] probably formed his idea of Tylor's total doctrine as he avowedly did his idea of that of Bentham, from knowledge of one detail; though he complained of other people who so conducted their criticism of his own writings."[17]

The reasons for this limitation of the understanding of other writers was due to his prejudices which made him an impatient reader, and due to what he called constitutional idleness, although likely that is hardly the correct diagnosis. He once stated that "I never could read books the cardinal principles of which I rejected."[18] This can be illustrated by the treatment that he

---

[14] *Ibid.,* Vol. I, pp. 201-202. See also *Life and Letters,* Vol. II, p. 309.

[15] Herbert Spencer, *First Principles,* 4th Edition, New York: D. Appleton and Co., 1897, p. 138.

[16] E.g., *Life,* p. 418.

[17] *Life,* p. 570. *History of Free Thought,* Vol. II, p. 356.

[18] Duncan, *op. cit.,* Vol. II, p. 312.

gave Comte's work on positive philosophy. "Being an impatient reader, especially when reading views from which I dissent, I did not go far."[19] And, it may be added, this kept him from going very far into the reasoning and the evidence that one would use in his book to support a position from which Spencer dissented.

Spencer's Attitude toward God

Like Darwin and Huxley, Spencer realized that evolution did not have to be associated with his own ontological views. Thus he wrote to W. D. Grounds, October 12, 1883, that "I quite agree with your statement that the general doctrine of Evolution is independent of these ontological views which I have associated with it; and I am not sorry to have this fact insisted upon. . . ."[20]

Spencer thought that there must be a First Cause, although it was unthinkable. "It is impossible to avoid making the assumption of self-existence somewhere; and whether that assumption be made nakedly, or under complicated disguises, it is equally vicious, equally unthinkable. . . . So that in fact, impossible as it is to think of the actual universe as self-existing, we do but multiply impossibilities of thought by every attempt we make to explain its existence."[21]

"We cannot think at all about the impressions which the external world produces on us, without thinking of them as caused; and we cannot carry out an inquiry concerning their causation, without inevitably committing ourselves to the hypothesis of a First Cause."[22] This, however, he thought involved us in other things that were inconceivable. There was a Power behind the Universe and which manifested itself through the Universe, but it was incomprehensible. ". . . the Power which the Universe manifests to us is utterly inscrutable."[23] Yet Spen-

---

19 *Ibid.*, Vol. II, p. 321.
20 *Ibid.*, Vol. I, p. 336. See also his *Autobiography*, Vol. I, pp. 170-171.
21 Spencer, *First Principles*, p. 37.
22 *Ibid.*, p. 38.
23 *Ibid.*, p. 48.

cer did seem to think that he knew enough about this power to say: "He, like every other man, may properly consider himself as one of the myriad agencies through whom works the Unknown Cause; and when the Unknown Cause produces in him a certain belief, he is thereby authorized to profess and act out that belief."[24] Even if the belief was belief in God?

Spencer, as other "inconceivables" mentioned by him show, could hardly mean that God did not exist because Spencer found Him inconceivable. For he affirmed the same thing of force and personality. "While then it is impossible to form any idea of Force in itself, it is equally impossible to comprehend its mode of exercise."[25] And speaking of personality he said: "So that the personality of which each is conscious, and of which the existence is to each a fact beyond all others the most certain, is yet a thing which cannot truly be known at all: knowledge of it is forbidden by the very nature of thought."[26] And yet, he did not deny their existence, nor did he affirm that therefore nothing could be known about them.

Spencer claimed that the belief in the reality of self was not reasonable, "... unavoidable as is this belief—established though it is not only by the assent of mankind at large, endorsed by divers philosophers, but by the suicide of the sceptical argument —it is yet a belief admitting of no justification by reason: nay, indeed, it is a belief which reason, when pressed for a distinct answer, rejects."[27] He, however, did not act as if he did not believe in the reality of his own existence.

Since Spencer was sceptical of the reality of his own existence, it is not surprising that he was agnostic concerning the existence of God. Like Darwin, he tried to avoid thinking about God, the Unknown. He wrote, June 26, 1895, to the Countess of Pembroke, that "It seems to me that our best course is to submit to the limitations imposed by the nature of our minds, and to

---

[24] *Ibid.*, p. 126.
[25] *Ibid.*, p. 63.
[26] *Ibid.*, p. 68.
[27] *Ibid.*, p. 67.

live as contentedly as we may in ignorance of that which lies behind things as we know them. My own feeling respecting the ultimate mystery is such that of late years I cannot even try to think of infinite space without some feeling of terror, so that I habitually shun the thought."[28] Thus his anti-supernaturalism so conquered his mind that he often refused even to think beyond it.[29]

Spencer was also like Darwin in that although he discredited human reason, he did not use this as an argument against the truth of evolution. And yet, if human reason is untrustworthy, if agnosticism is the proper attitude, why should one put any confidence in the hypothesis of evolution, which hypothesis is a product of human reasoning?

This chapter has shown that Spencer also had been so influenced by the idea of uniformity that he was unwilling to consider the idea of a supernatural creation. Thus he accepted a hypothesis of evolution, and was very gratified when he thought that Darwin had proved the hypothesis.[30]

Now let us see if such was also true of T. H. Huxley.

---

[28] *Life and Letters of Herbert Spencer,* Vol. II, p. 83.

[29] For an examination of the agnosticism of Spencer, see William Arthur, *Religion without God,* London: Bemrose and Sons, 1888. In Duncan, *Life and Letters of Herbert Spencer,* Vol. I, p. 105, Spencer denied that he was an atheist and maintained that "the existence of a Deity can neither be proved nor disproved."

[30] *An Autobiography,* Vol. II, pp. 57-58.

CHAPTER V

THOMAS HENRY HUXLEY

Multitudes were led to accept the theory of evolution because T. H. Huxley championed it so vigorously and so long. It was Huxley, by his ability to turn an argument or a phrase, who was a major factor in turning the tide in favor of evolution in the Oxford Meeting of 1860.[1]

Huxley himself pointed out that he was ready from the very beginning to do battle for Darwin. After reading the *Origin of Species* he wrote to Darwin on November 23, 1859, that: "I trust you will not allow yourself to be in any way disgusted or annoyed by the considerable abuse and misrepresentation which, unless I greatly mistake, is in store for you. Depend upon it, you have earned the lasting gratitude of all thoughtful men. And as to the curs which will bark and yelp, you must recollect that some of your friends, at any rate, are endowed with an amount of combativeness which (though you have often and justly rebuked it) may stand you in good stead.

"I am sharpening up my claws and beak in readiness."[2]

" 'I am Darwin's bull-dog,' he said, and the *Quarterly Reviewer's* treatment of Darwin, 'alike unjust and unbecoming,' provoked him into immediate action."[3] Darwin wrote to Huxley that "The pendulum is now swinging against our side, but I feel positive it will soon swing the other way; and no mortal man

---

[1] Leonard Huxley, *Life and Letters of Thomas Henry Huxley*, New York: The Macmillan Co., 1903, Vol. I, pp. 259-274.

[2] *Ibid.*, Vol. I, p. 254. Also *Life and Letters of Charles Darwin*, Vol. II, p. 27.

[3] Leonard Huxley, *op. cit.*, Vol. II, p. 62.

will do half as much as you in giving it a start in the right direction, as you did at the first commencement."[4]

Of others who paid tribute to Huxley's influence we shall quote only his son, Bateson, and Lord Kelvin. His son wrote that "Under the suggestive power of the *Origin of Species* all these scattered studies fell suddenly into due rank and order; the philosophic unity he had so long been seeking inspired his thought with tenfold vigour, and the battle at Oxford in defence of the new hypothesis first brought him before the public eye as one who not only had the courage of his convictions when attacked, but could, and more, would, carry the war effectively into the enemy's country. And for the next ten years he was commonly identified with the championship of the most unpopular view of the time; a fighter, an assailant of long-established fallacies, he was too often considered a mere iconoclast, a subverter of every other well-rooted institution, theological, educational, or moral."[5]

When Lord Kelvin, President of the Royal Society, awarded to Huxley in 1894 the Darwin Medal, he paid high tribute to Huxley's work in the spread of Darwinism. "To the world at large, perhaps, Mr. Huxley's share of moulding the thesis of *Natural Selection* is less well known than is his bold unwearied exposition and defence of it after it had been made public. And, indeed, a speculative trifler, revelling in the problems of the 'might have been,' would find a congenial theme in the inquiry how soon what we now call 'Darwinism' would have met with the acceptance with which it has met, and gained the power which it has gained, had it not been for the brilliant advocacy with which in its early days it was expounded to all classes of men."[6]

The geneticist Bateson paid him this tribute: "To the world, scientific as well as lay, Huxley is chiefly famous as the champion of evolutionary doctrine, whose vigorous and skillful ad-

---

[4] *Ibid.,* Vol. II, p. 64.
[5] *Ibid.,* Vol. II, pp. 1-2.
[6] *Ibid.,* Vol. I, p. 301.

vocacy counted for so much in obtaining the favourable verdict of the public."[7]

It is thus very important to ask: What influenced Huxley, who influenced so many, to accept the doctrine of evolution? It will be shown that it was, as in the case of the others, his anti-supernatural bias. And thus it was that a vigorous man with a strong bias persuaded multitudes that the hypothesis of evolution was a scientific fact.

Uniformity First

Huxley early cultivated an attitude of disrespect for authority, which evidently soon extended to disrespect for authority as manifested in God and religion. In writing to Spencer, November 25, 1886, after reading some proof sheets of Spencer's *Autobiography,* Huxley said: "Another point which has interested me immensely is the curious similarity to many recollections of my own boyish nature which I find, especially in the matter of demanding a reason for things and having no respect for authority.

"But I was more docile, and could remember anything I had a mind to learn, whether it was rational or irrational, only in the latter case I hadn't the mind."[8]

Huxley also early fell under the influence of Hutton's uniformity. "Not satisfied with the ordinary length of the day, he used, when a boy of twelve, to light his candle before dawn, pin a blanket around his shoulders, and sit up in bed to read Hutton's *Geology.*"[9]

Years later in writing about the reception of the *Origin of Species* he summed up some of the characteristics of his thinking at the time that the *Origin* appeared. "I think I must have read the *Vestiges* before I left England in 1846; but, if I did, the book made very little impression upon me, and I was not brought

---

[7] "Huxley and Evolution," in Beatrice Bateson, *William Bateson, F. R. S. Naturalist.* Cambridge: At the University Press, 1928, p. 460.

[8] Leonard Huxley, *op. cit.,* Vol. II, p. 471.

[9] *Ibid.,* Vol. I, p. 8.

into serious contact with the 'Species' question until after 1850. At that time, I had long done with the Pentateuchal cosmogony, which had been impressed upon my childish understanding as Divine truth, with all the authority of parents and instructors, and from which it had cost me many a struggle to get free. But my mind was unbiased in respect of any doctrine which presented itself, if it professed to be based on purely philosophical and scientific reasoning. It seemed to me then (as it does now) that 'creation,' in the ordinary sense of the word, is perfectly conceivable. I find no difficulty in conceiving that, at some former period, this universe was not in existence; and that it made its appearance in six days (or instantaneously, if that is preferred), in consequence of the volition of some pre-existing Being. Then, as now, the so-called *a priori* arguments against Theism, and, given a Deity, against the possibility of creative acts, appeared to me to be devoid of reasonable foundation. I had not then, and I have not now, the smallest *a priori* objection to raise to the account of the creation of animals and plants given in *Paradise Lost,* in which Milton so vividly embodies the natural sense of Genesis. Far be it from me to say that it is untrue because it is impossible. I confine myself to what must be regarded as a modest and reasonable request for some particle of evidence that the existing species of animals and plants did originate in that way, as a condition of my belief in a statement which appears to me to be highly improbable.

"And, by way of being perfectly fair, I had exactly the same answer to give to the evolutionists of 1851-58. Within the ranks of the biologists, at that time, I met with nobody, except Dr. Grant of University College, who had a word to say for Evolution—and his advocacy was not calculated to advance the cause. Outside these ranks, the only person known to me whose knowledge and capacity compelled respect, and who was, at the same time, a thorough-going evolutionist, was Mr. Herbert Spencer, whose acquaintance I made, I think, in 1852, and then entered into the bonds of a friendship which, I am happy to think, has known no interruption. Many and prolonged were the battles we fought on this topic. But even my friend's rare dialectic skill and copiousness of apt illustration could not drive me from my agnostic position. I took my stand upon two grounds:— Firstly, that up to that time, the evidence in favour of transmutation was wholly insufficient; and secondly, that no suggestion respecting the causes of transmutation assumed, which had been made, was in any way adequate to explain the phenomena. Looking back at the state of knowledge at that time, I really do not see that any other conclusion was justifiable.

"In those days I had never even heard of Treviranus' *Biologie*. However, I had studied Lamarck attentively, and I had read the *Vestiges* with due care; but neither of them afforded me any good ground for changing my negative and critical attitude. As for the *Vestiges*, I confess that the book simply irritated me by the prodigious ignorance and thoroughly unscientific habit of mind manifested by the writer. If it had any influence on me at all, it set me against Evolution; and the only review I ever have qualms of conscience about, on the ground of needless savagery, is one I wrote on the *Vestiges* while under that influence. . . .

"But, by a curious irony of fate, the same influence which led me to put as little faith in modern speculations on this subject as in the venerable traditions recorded in the first two chapters of Genesis, was perhaps more potent than any other in keeping alive a sort of pious conviction that Evolution, after all, would turn out true. I have recently read afresh the first edition of the *Principles of Geology;* and when I consider that this remarkable book had been nearly thirty years in everybody's hands, and that it brings home to any reader of ordinary intelligence a great principle and a great fact,—the principle that the past must be explained by the present, unless good cause be shown to the contrary; and the fact that so far as our knowledge of the past history of life on our globe goes, no such cause can be shown,—I cannot but believe that Lyell, for others, as for myself, was the chief agent in smoothing the road for Darwin. For consistent uniformitarianism postulates Evolution as much in the organic as in the inorganic world. The origin of a new species by other than ordinary agencies would be a vastly greater 'catastrophe' than any of those which Lyell successfully eliminated from sober geological speculation.

"As I have already said, I imagine that most of those of my contemporaries who thought seriously about the matter were very much in my own state of mind—inclined to say to both Mosaists and Evolutionists, 'a plague on both your houses!' and disposed to turn aside from an interminable and apparently fruitless discussion, to labour in the fertile fields of ascertainable fact. And I may therefore suppose that the publication of the Darwin and Wallace paper in 1858, and still more that of the 'Origin' in 1859, had the effect upon them of the flash of light which, to a man who has lost himself on a dark night, suddenly reveals a road which, whether it takes him straight home or not, certainly goes his way. That which we were looking for, and could not find, was a hypothesis respecting the origin of known organic forms which assumed the operation of no causes but such as could be proved to be actually at work. We wanted, not to pin our faith to that or any other speculation, but to get hold of clear and definite conceptions which could be brought face to face with facts and have their validity tested. The 'Origin' provided us with the working hypothesis we sought. Moreover, it did the immense service of freeing us for ever from the dilemma—Refuse to accept the creation hypothesis, and what have you to propose that can be accepted by any cautious reasoner? In 1857 I had no answer ready, and I do not think that any one else had. A year later we reproached ourselves with dulness for being perplexed with such an inquiry. My reflection, when I first made myself master of the central idea of the 'Origin' was, 'How extremely stupid not to have thought of that!' I suppose that Columbus' companions said much the same when he made the egg stand on end. The facts of variability, of the struggle of existence, of adaptation to conditions, were notorious enough; but none of us had suspected that the road to the heart of the species problem lay through them, until Darwin and Wallace dispelled the darkness, and the beacon-fire of the 'Origin' guided the benighted.

"Whether the particular shape which with the doctrine of Evolution, as applied to the organic world, took in Darwin's hands, would prove to be final or not, was to me a matter of indifference. In my earliest criticisms of the 'Origin' I ventured to point out that its logical foundation was insecure so long as experiments in selective breeding had not produced varieties which were more or less infertile; and that insecurity remains up to the present time. But, with any and every critical doubt which my sceptical ingenuity could suggest, the Darwinian hypothesis remained incomparably more probable than the creation hypothesis. And if we had none of us been able to discern the paramount significance of some of the most patent and notorious of natural facts, until they were, so to speak, thrust under our noses, what force remained in the dilemma—creation or nothing? It was obvious that hereafter the probability would be immensely greater, that the links of natural causation were hidden from our purblind eyes, than that natural causation should be incompetent to produce all the phenomena of nature. The only rational course for those who had no other object than the attainment of truth was to accept 'Darwinism' as a working hypothesis and see what could be made of it. Either it would prove its capacity to elucidate the facts of organic life, or it would break down under the strain. This was surely the dictate of common sense, and, for once, common sense carried the day."[10]

These statements from T. H. Huxley are worthy of comment.

Supernatural First Rejected

With Huxley, as with Spencer, it is instructive to notice that his rejection of the supernatural came before his acceptance of evolution. The evidence for evolution did not lead him to abandon creation; instead his abandonment of creation led him finally to accept the hypothesis of evolution. As his son Leonard wrote, "Before this time [the publication of *Origin of Species*

---

[10] *Ibid.,* Vol. I, pp. 241-244.

in 1859, J.D.B.], he had taken up a thoroughly agnostic attitude with regard to the species question, for he could not accept the creational theory, yet sought in vain among the transmutationists for any cause adequate to produce transmutation."[11] And yet, Huxley was bound to accept some hypothesis of evolution if he accepted any position at all concerning the origin of things, since he had already rejected creation. Furthermore, as he also pointed out, he needed some explanation in order to avoid the dilemma into which the creationists put him.[12]

Inadequate Grounds

Huxley for a time had maintained an agnostic position (i.e., that he did not know) toward evolution. "I took my stand upon two grounds:—Firstly, that up to that time, the evidence in favour of transmutation was wholly insufficient; and secondly, that no suggestion respecting the causes of transmutation assumed, which had been made, was in any way adequate to explain the phenomena. Looking back at the state of knowledge at that time, I really do not see that any other conclusion was justifiable."[13] He should have continued to have stood on these two grounds for evolutionists generally acknowledge today that Darwin's theory itself does not explain evolution. It tried to explain the survival of the fittest but not the arrival of the fittest. Furthermore, anyone who has read the *Origin of Species* should be able to see that a work cannot scientifically establish a hypothesis if it has to be bolstered up with hundreds of "perhapses," or other expressions of uncertainty, and when it assumes or jumps over evidences that are vital to its proof.

The Wish

In rejecting Genesis and holding to uniformity there was kept alive "a sort of pious conviction that Evolution, after all, would turn out true."[14] This confession of Huxley again gives

---

[11] *Ibid.*, Vol. I, p. 239.
[12] *Ibid.*, Vol. I, p. 246.
[13] *Ibid.*, Vol. I, p. 242.
[14] *Ibid.*, Vol. I, p. 243.

us an insight into his state of mind. With no alternative hypothesis to creation, with a rejection of creation, with a belief in uniformity, and with an agnostic attitude toward evolution, there still lurked the idea in Huxley's mind that evolution would be shown to be true. And that he must have wished it to be true is shown by the fact that it would, if true, let him out of an embarrassing dilemma.[15] "The *'Origin'* provided us with the working hypothesis we sought."[16]

The *Origin of Species* found Huxley in a very receptive frame of mind for the additional reason that it gave him a stick with which to beat the creationists as well as to let him out of an embarrassing dilemma.[17]

Huxley was very glad to accept evolution since he had been looking for some "acceptable" hypothesis of evolution due to his prior acceptance of the theory of uniformity.

Uniformity

Huxley had accepted the theory of uniformity as a "great principle and a great fact." "I cannot but believe," he said, "that Lyell, for others, as for myself, was the chief agent in smoothing the road for Darwin. For consistent uniformitarianism postulates Evolution as much in the organic as in the inorganic world."[18] "But, with any and every critical doubt which my sceptical ingenuity could suggest, the Darwinian hypothesis remained incomparably more probable than the creation hypothesis."[19] Yes, it was indeed the only plausible thing if uniformity were true and creation rejected. The "whole theory crumbles to pieces if the uniformity and regularity of natural causation for illimitable past ages is denied."[20]

---

15 *Ibid.*, Vol. I, p. 246
16 *Ibid.*, Vol. I, pp. 245-246.
17 *Ibid.*, Vol. I, p. 246.
18 *Ibid.*, Vol. I, p. 243.
19 *Ibid.*, Vol. I, p. 246.
20 T. H. Huxley in *Life and Letters of Charles Darwin,* Vol. I, p. 553.

Not Proved

"I by no means," wrote Huxley to Charles Lyell (June 25, 1859), "suppose that the transmutation hypothesis is proven or anything like it. But I view it as a powerful instrument of research. Follow it out, and it will lead us somewhere; while the other notion is like all the modifications of 'final causation,' a barren virgin.

"And I would very strongly urge upon you that it is the logical development of Uniformitarianism, and that its adoption would harmonise the spirit of Paleontology with that of Physical Geology."[21] Huxley was wrong in assuming that belief in creation must of necessity frustrate research. Believers in God have made more contributions to science than have believers in materialism, or agnostics. This is evident from the fact that most scientists hold to some kind of theism. Furthermore, although such a theory might harmonize what he considered to be the spirit of Paleontology with that of Physical Geology, it was not necessary to investigation in these fields. It is not necessary since the scientists in both fields can deal with the facts in these fields without accepting evolution.

However, it is not our purpose here to argue with Huxley's statements, but rather to present them that it may be evident that he accepted evolution as a deduction from a previous theory —that of uniformity.[22]

Huxley Actually Ruled Out Creation

Huxley considered men fools for believing in "...the myths in Genesis. But my sole point is to get the people who persist in regarding them as statements of fact to understand that they are fools."[23] However, he also wrote: "True, one must believe in a beginning somewhere, but science consists in not believing the having reached that beginning before one is forced to do so.

---

[21] Leonard Huxley, *op. cit.*, Vol. I, p. 252.
[22] See also *Life and Letters of Charles Darwin,* Vol. I, pp. 543, 545, 547, 553.
[23] Leonard Huxley, *op. cit.*, Vol. II, p. 429. See also p. 430.

"It is wholly impossible to prove that any phenomenon whatsoever is not produced by the interposition of some unknown cause. But philosophy has prospered exactly as it has disregarded such possibilities, and has endeavored to resolve every event by ordinary reasoning."[24]

Like Darwin, Huxley stated the creationist position in a way in which a creationist would be foolish to state it, and which, so far as the authors' knowledge goes, no creationist has stated it. In fact, Huxley's statement of the creationist position and its difficulty is his position and his difficulty and not that of the creationist. In writing to Charles Lyell, June 25, 1859, he asked: "How much evidence would you require to believe that there was a time when stones fell upwards, or granite made itself by a spontaneous rearrangement of the elementary particles of clay and sand? And yet the difficulties in the way of these beliefs are as nothing compared to those which you would have to overcome in believing that complex organic beings made themselves (for that is what creation comes to in scientific language) out of inorganic matter."[25]

Huxley's position, not ours, is the one that assumes that complex organic beings made themselves for he does not accept the Creator whose creative power brought about the creation. He thinks that matter kept shaking until it shook out complex organic beings. Huxley to be consistent had to believe that matter made or created these first creatures.

Huxley continued by saying: "I know it will be said that even on the transmutation theory, the first organic being must have made itself. But there is as much difference between supposing the passage of inorganic matter into an *amoeba*, e.g., and into an *Elephant*, as there is between supposing that Portland stone might have built itself up into St. Paul's, and believing

---

[24] *Ibid.*, Vol. I, p. 251; see also p. 252. T. H. Huxley, in letter to Sir Charles Lyell, June 25, 1859. For some reasons for believing in God, see James D. Bales, *Communism: Its Faith and Fallacies,* Grand Rapids 6, Michigan: Baker Book House, 1962.

[25] Leonard Huxley, *op. cit.,* Vol. I, p. 251.

that the Giant's Causeway may have come about by natural causes."[26] This statement does not, however, state the difficulty of the creationist but of the evolutionist.

Huxley and Belief in God

Huxley regarded himself as an agnostic.[27] Atheism was untenable. ".... to my mind, atheism is, on purely philosophical grounds, untenable. That there is no evidence of the existence of such a being as the God of the theologians is true enough; but strictly scientific reasoning can take us no further. Where we know nothing we can neither affirm nor deny with propriety."[28]

Huxley did not think that there were any *a priori* arguments against God, nor did he contend that evolution had undermined every type of argument from design for the existence of God. Concerning *a priori* arguments against God he said: "Then, as now, the so-called *a priori* arguments against Theism; and, given a Deity, against the possibility of creative acts, appeared to me to be devoid of reasonable foundation."[29]

Concerning design Huxley wrote: "A second very common objection to Mr. Darwin's views was (and is), that they abolish Teleology, and eviscerate the argument from design. It is nearly twenty years since I ventured to offer some remarks on this subject, and as my arguments have as yet received no refutation, I hope I may be excused for reproducing them. I observed, 'that the doctrine of Evolution is the most formidable opponent of all the commoner and coarser forms of Teleology. But perhaps the most remarkable service to the Philosophy of Biology rendered by Mr. Darwin is the reconciliation of Teleology and Morphology, and the explanation of the facts of both, which his views offer. The teleology which supposes that the eye, such as we see it in man, or one of the higher vertebrata, was made

---

[26] *Ibid.,* Vol. I, p. 251.
[27] *Ibid.,* Vol. III, p. 97.
[28] *Ibid.,* Vol. III, pp. 18-19.
[29] *Life and Letters of Charles Darwin,* Vol. I, p. 541. See also pp. 554-556.

with the precise structure it exhibits, for the purpose of ena-
bling the animal which possesses it to see, has undoubtedly received its death-blow. Nevertheless, it is necessary to remember
that there is a wider teleology which is not touched by the doctrine of Evolution, but is actually based upon the fundamental
proposition of Evolution. This proposition is that the whole
world, living and not living, is the result of the mutual inter-
action, according to definite laws, of the forces (powers) possessed by the molecules of which the primitive nebulosity of the
universe was composed. If this be true, it is no less certain that
the existing world lay potentially in the cosmic vapour,   and
that a sufficient intelligence could, from a knowledge of the
properties of the molecules of that vapour, have predicted, say
the state of the fauna of Britain in 1869, with as much certainty
as one can say what will happen to the vapour of the breath
on a cold winter's day. . . .

"'. . . The teleological and the mechanical views of nature
are not, necessarily, mutually exclusive. On the contrary, the
more purely a mechanist the speculator is, the more firmly
does he assume a primordial molecular arrangement of which
all the phenomena of the universe are the consequences, and
the more completely is he thereby at the mercy of the teleologists,
who can always defy him to disprove that this primordial molecular arrangement was not intended to evolve the phenomena of
the universe.'

"The acute champion of Teleology, Paley, saw no difficulty
in admitting that the 'production of things' may be the result
of trains of mechanical dispositions fixed beforehand by intelligent appointment and kept in action by a power at the centre
('Natural Theology'). . . .

"Having got rid of the belief in chance and the disbelief in
design, as in no sense appurtenances of Evolution, the third
libel upon that doctrine, that it is anti-theistic, might perhaps
be left to shift for itself. But the persistence with which many
people refuse to draw the plainest consequences from the propositions they profess to accept, renders it advisable to remark

that the doctrine of Evolution is neither Anti-theistic nor The-istic. It simply has no more to do with Theism than the first book of Euclid has. It is quite certain that a normal fresh-laid egg contains neither cock nor hen; and it is also as certain as any proposition in physics or morals, that if such an egg is kept under proper conditions for three weeks, a cock or hen chicken will be found in it. It is also quite certain that if the shell were transparent we should be able to watch the formation of the young fowl, day by day, by a process of evolution, from a microscopic cellular germ to its full size and complication of structure. Therefore Evolution, in the strictest sense, is actually going on in this and analogous millions and millions of instances, wherever living creatures exist. Therefore, to borrow an argu-ment from Butler, as that which now happens must be consis-tent with the attributes of the Deity, if such a Being exists, Evolution must be consistent with those attributes. And, if so, the evolution of the universe, which is neither more nor less explicable than that of a chicken, must also be consistent with them. The doctrine of Evolution, therefore, does not even come into contact with Theism, considered as a philosophical doctrine. That with which it does collide, and with which it is absolutely inconsistent, is the conception of creation, which theological speculators have based upon the history narrated in the opening of the book of Genesis.

"There is a great deal of talk and not a little lamentation about the so-called religious difficulties which physical science has created. In theological science, as a matter of fact, it has created none. Not a solitary problem presents itself to the philosophical Theist, at the present day, which has not existed from the time that philosophers began to think out the logical grounds and the logical consequences of Theism."[30]

In some sense, Huxley thought, the cosmic process was ra-tional.[31] It was "utter bosh" that Darwinism was atheistic.[32] He

---

[30] *Ibid.*, Vol. I, pp. 554-556.
[31] Leonard Huxley, *op. cit.*, Vol. III, p. 218.
[32] *Ibid.*, Vol. II, p. 467. Oct. 20, 1886.

disliked materialism more than he did spiritualism, and thought that atheism was a greater absurdity than theism.[33]

That Huxley did not think that evolution had to be anti-theistic is evident also from the advice which, as he wrote to Herbert Spencer, June 15, 1875, he gave to one of his classes: "I have a class of 353, and instruct them in dry facts—particularly warning them to keep free of the infidel speculations which are current under the name of evolution."[34]

Huxley even put forward a line of argument which led to a theistic conclusion. He wrote: "The student of nature, who starts from the axiom of the universality of the law of causation, cannot refuse to admit an eternal existence; if he admits the conservation of energy, he cannot deny the possibility of an eternal energy; if he admits the existence of immaterial phenomena in the form of consciousness, he must admit the possibility, at any rate, of an eternal series of such phenomena; and, if his studies have not been barren of the best fruit of the investigation of nature, he will have enough sense to see that when Spinoza says, 'Per deum intelligo ens absolute infinitum, hoc est substantiam constantem infinitis attributis,'[35] the God so conceived is one that only a very great fool would deny, even in his heart. Physical science is as little Atheistic as it is materialistic."[36]

Huxley admitted that consciousness exists. "I understand the main tenet of Materialism to be that there is nothing in the universe but matter and force; and that all the phenomena of

---

[33] *Ibid.,* Vol. I, pp. 350, 323-324, 431-432; Vol. II, pp. 133, 470, 195.

[34] *Ibid.,* Vol. II, p. 180. Huxley did not want organized religion destroyed (*ibid.,* Vol. I, p. 320). He thought that the Bible should be taught for its moral value (Vol. II, pp. 26, 31-37, 273; Vol. III, p. 214).

[35] "By the word God I understand a being absolutely infinite, that is, a constant (self-existing) substance with infinite attributes." Translation by Dr. W. M. Green.

[36] *Evolution and Ethics and Other Essays.* New York: D. Appleton and Co., 1896, p. 140.

nature are explicable by deduction from the properties assignable to these two primitive factors. That great champion of Materialism whom Mr. Lilly appears to consider to be an authority in physical science, Dr. Buchner, embodies this article of faith on his title-page. *Kraft and Stoff*—force and matter—are paraded as the Alpha and Omega of existence. This I apprehend is the fundamental article of the faith materialistic; and whosoever does not hold it is condemned by the more zealous of the persuasion (as I have some reason to know) to the Inferno appointed for fools or hypocrites. But all this I heartily disbelieve; and at the risk of being charged with wearisome repetition of an old story, I will briefly give my reasons for persisting in my infidelity. In the first place, as I have already hinted, it seems to me pretty plain that there is a third thing in the universe, to wit, consciousness, which, in the hardness of my heart or head, I cannot see to be matter, or force, or any conceivable modification of either, however intimately the manifestations of the phenomena of consciousness may be connected with the phenomena known as matter and force. In the second place, the arguments used by Descartes and Berkeley to show that our certain knowledge does not extend beyond our states of consciousness, appear to me to be as irrefragable now as they did when I first became acquainted with them some half-century ago. All the materialistic writers I know of who have tried to bite that file have simply broken their teeth. But, if this is true, our one certainty is the existence of the mental world, and that of *Kraft and Stoff* falls into the rank of, at best, a highly probable hypothesis."[37]

Huxley also realized that consciousness had not always existed on this earth. Being unable to believe that consciousness is a form or modification of matter and force, he could hardly have believed that it was produced by them. For in that case it would have been a form or modification of them. Regardless of how you arrange matter and force they remain matter and force; that is, unless life, which is not a mere arrangement of matter and

---

[37] *Ibid.*, pp. 129-130.

force, moulds matter into certain forms. When to this is added his adherence to the law of causation, it would seem that Huxley had to look to some superhuman and supernatural form of consciousness in order to find the cause for consciousness on this earth. We say superhuman, for surely no one believes that man came from some other planet and thus brought consciousness here. We say supernatural, for consciousness, Huxley seems to grant, was not produced by natural forces on this earth as a modification or form of matter or force.

Spontaneous Generation

Huxley viewed spontaneous generation as "a necessary corollary from Darwin's views if legitimately carried out, and I think Owen smites him (Darwin) fairly for taking refuge in 'Pentateuchal' phraseology when he ought to have done one of two things—(a) give up the problems, (b) admit the necessity of spontaneous generation. It is the very passage in Darwin's book to which, as he knows right well, I have always strongly objected."[38]

The spontaneous generation of life, from non-life, was another conclusion that Huxley accepted because it was inherent in his uniformitarianism. With him it was an act of faith, of philosophic faith, as he called it. In his Presidential Address at the British Association for 1870 "he discussed the rival theories of spontaneous generation and the universal derivation of life from precedent life, and professed his belief," said his son, "as an act of philosophic faith, that at some remote period, life had arisen out of inanimate matter, though there was no evidence that anything of the sort had occurred recently, the germ theory explaining many supposed cases of spontaneous generation."[39] It is interesting to note that Huxley called on faith in this connection, although his son termed it philosophic faith. It would have been more consistent with his agnosticism to have said that as all the evidence is against it, whether it took place

---

[38] Leonard Huxley, *op. cit.*, Vol. I, p. 352.
[39] *Ibid.*, Vol. II, pp. 15-16.

at one time was a matter concerning which he would have to be agnostic.

The fact of the matter is that one of the best established principles in science is that life comes from life. There is no known exception.[40]

Evolution a Hypothesis

In a letter to Lyell, June 25, 1859, Huxley stated that "I by no means suppose that the transmutation hypothesis is proven or anything like it."[41] In 1862 it was "a working hypothesis like other scientific generalisations, 'subject to the production of proof that physiological species may be produced by selective breeding. . . .' "[42] In 1876 he claimed that it had been demonstrated, but he did not appeal to a demonstration by selective breeding, which he had said in 1862 was essential to its proof. What he meant "by demonstrative evidence of evolution" was that: "An inductive hypothesis is said to be demonstrated when the facts are shown to be in entire accordance with it."[43] In 1893 Huxley still held the conviction that the logical foundation of natural selection had not been completed. In the Preface to *Darwiniana* he wrote: "As I have said in the seventh essay, the fact of evolution is to my mind sufficiently evidenced by palaeontology; and I remain of the opinion expressed in the second, that until selective breeding is definitely proved to give rise to varieties infertile with one another, the logical foundation of the theory of natural selection is incomplete."[44]

There are some who maintain that to the very end Huxley did not become a convinced believer in natural selection. "Sir Edward Poulton of Oxford University has said, 'The attitude

---

[40] On the origin of life see Robert E. D. Clark, *Darwin: Before and After,* chapters I-III. Also James D. Bales, *Communism: Its Faith and Fallacies,* Grand Rapids, Michigan: Baker Book House, 1962, pp. 58-65. We hope to devote an entire monograph to this question.

[41] Leonard Huxley, *op. cit.,* Vol. I, p. 252.

[42] *Ibid.,* Vol. I, p. 279.

[43] *Ibid.,* Vol. II, p. 213.

[44] *Ibid.,* Vol. II, p. vi.

of Huxley towards Natural Selection was...remarkable and unusual. Although no one strove so nobly and against such odds in its defence from unfair attack, although no one ever fought the battle of science with more complete success, Huxley was at no time a convinced believer in the theory he protected.' "[45]

The reason that Huxley continued to hold to Darwinism in spite of its difficulties was that with him it was Darwinism or nothing. "I really believe that the alternative is either Darwinism or nothing, for I do not know of any rational conception or theory of the organic universe which has any scientific position at all beside Mr. Darwin's.... Whatever may be the objections to his views, certainly all other theories are absolutely out of court." "But you must recollect that when I say I think it is either Mr. Darwin's hypothesis or nothing; that either we must take his view, or look upon the whole of organic nature as an enigma, the meaning of which is wholly hidden from us; you must understand that I mean that I accept it provisionally, in exactly the same way as I accept any other hypothesis."[46]

This chapter has made it evident that Huxley accepted the theory of evolution not because he believed that it was scientifically established but because he was a uniformitarian who had ruled out creation and had to accept evolution or nothing, and he did not want to accept nothing.

---

[45] *Essays on Evolution,* Oxford: The Clarendon Press, 1908, p. 193. Quoted in *Evolution,* Toronto, Ontario, Canada: International Christian Crusade, p. 39.

[46] T. H. Huxley, *Darwiniana.* New York: D. Appleton and Co., 1896, pp. 467, 468. Statement made in 1863 and reprinted in 1893 by Huxley.

CHAPTER VI

ALFRED RUSSELL WALLACE

Alfred Russell Wallace came to the theory of natural selection, or survival of the fittest, independently of, but not before, Darwin.[1] December 28, 1847 it was pointed out that the book on *Vestiges of the Natural History of Creation* appealed to him and that he thought it established the conception of evolution through natural law.[2] Like Darwin, Wallace was also influenced by Malthus.[3]

He became an unbeliever in the Bible before he became an evolutionist. However, he did retain a spiritualistic view of life and a type of theism.[4]

Lyell and the doctrine of uniformity also led Wallace to evolution, as is implied in a letter to his brother-in-law, Thomas Sims, on March 15, 1861. In writing concerning Darwin's *Origin of Species,* Wallace maintained that: "The evidence for the production of the organic world by the simple laws of inheritance is exactly of the same nature as that for the production of the present surface of the earth—hills and valleys, plains, rocks, strata, volcanoes, and all their fossil remains—by the slow and natural action of natural causes now in operation. The mind that will ultimately reject Darwin must (to be consistent) reject Lyell also. The same arguments of apparent sta-

---

[1] James Marchant, *Alfred Russell Wallace: Letters and Reminiscences.* London: Cassell and Co., Ltd., 1916, Vol. I, pp. 170-174; *Life and Letters of Charles Darwin,* Vol. I, pp. 472-482.

[2] Marchant, *op. cit.,* Vol. I, pp. 91-92.

[3] *Ibid.,* Vol. I, pp. 111, 113, 116.

[4] *My Life,* Vol. I, pp. 78, 87ff., 266-268; Marchant, *op cit.,* Vol. I, pp. 66-67, 341-342.

bility which are thought to disprove that organic species can change will also disprove any change in the inorganic world, and you must believe with your forefathers that each hill and each river, each inland lake and continent, were created as they stand, with their various strata and their various fossils— all appearances and arguments to the contrary notwithstanding."[5]

There are several fallacies in this paragraph which we shall notice.

Proving the Wrong Conclusions

By observation men can see at work now natural processes that can carve out hills, open up valleys, build up plains, produce volcanoes, and so forth. But because this is done it does not follow that life and all the various forms of life are the result of the slow, non-directed working of natural processes. In the *first* place, one is dealing with an entirely different realm —in its essence—when he is dealing with the realm of life, than when he is dealing with the realm of non-living matter. In the *second* place, where is the proof that natural forces can create life and produce all the various forms of life existing today? There is no such proof, although there is proof that natural forces can carve out a canyon.

To prove that there is some change in the world of living things, proves that there is some change; but it does not prove that the unlimited change postulated by evolution has taken place. Thus even if it is true that some so-called species are not as stable as some have thought, it does not follow that there is the unlimited change that the hypothesis of evolution demands. To take cases of limited change and prove by these cases unlimited change, is to prove the wrong conclusions. Limited change proves limited change.

Appeal to Consequences

One can legitimately use the appeal to consequences when he shows that the consequences of a position result in contra-

---

[5] Marchant, *op. cit.*, Vol. I, p. 78.

dictions to well-established facts. Wallace, however, argues that one must—to be consistent—reject Lyell, if he rejects Darwin. But that is not an argument for accepting Darwinism, unless one has first shown that Lyell ought to be accepted. Wallace has reference to Lyell's doctrine of uniformity, which taught that all changes in the inorganic world can be accounted for by the operating of laws that are even now operating. Present laws explain all past changes in the inorganic world, according to this view. Darwin simply extended this view to the world of living things, and maintained that laws even now operating account for all the various forms of life that surround us. Thus, creation was not a supernatural act by God, but a natural act of the laws of nature.

It is true that a view of nature that leaves out God and claims that all things are possible without God, would prepare one's mind for the reception of a similar view concerning the world of living things. And yet, to prove uniformity for the inorganic world would not prove the same doctrine for the organic world. But even if rejecting Darwin involved—as a necessary consequence—the rejection of Lyell it would not mean that one should hesitate to reject Darwin. If there is sufficient reason to reject Darwin, it would be right to reject Lyell—if Darwinism followed as a necessary consequence from Lyell. Thus, even if Wallace proved that one must reject Lyell, to be consistent, once he has rejected Darwin, it would not prove that one ought to accept Darwin. Thus, to appeal to this supposed consequence of rejecting Darwin does not in any way make a sound argument for accepting Darwin, if rejecting him does not involve us in rejecting any known fact. There is, of course, a vast difference between rejecting a fact and rejecting someone's interpretation of the fact.

Caricaturing the Opposition

It is possible to so misstate the position of the opposition that it is obviously ridiculous and unsound. Then one can imply that if the ridiculous position of the opposition is rejected, one's own must be accepted. Wallace states the creationist position in

such a way as to make it ridiculous. And since one would obviously reject it, he implies that the evolutionist's position is the only
alternative and therefore ought to be accepted. But what creationist has ever held the extreme view that Wallace attributes to them?
The authors have heard of only one. What creationist believes
that the position in which everything stands today—the hills
and rivers, etc.—is exactly the position in which they were
created by God in the beginning?

There are creationists who believe that the flood of Noah's
day was universal, and that many changes were introduced by
it.[6] Certainly they do not believe that things are now standing
just as they were the moment after they were created. Furthermore, all creationists believe that natural laws have been in
operation, and have been acting on the world, since they were
established in the beginning of the world. Creationists realize
that these natural forces have wrought changes. Wallace should
not have charged them with the view that he represented them
as holding. The fact that he charged them with such an obviously unsound position indicates that there was a deep bias
operating in his mind against the position of the creationist.
If it does not show this, then it manifests an amazing ignorance
of the informed creationist's view; an ignorance that is hard
to account for, since it should be obvious that informed creationists could not believe in such an absurd position and still
hold to their belief that God has natural laws. Wallace certainly knew that creationists believed, for example, that there
is sunshine and rain, earthquakes, and other natural forces that
make an impression on the physical world.

This is the same fallacy that Darwin used to disprove design,
and which he was so anxious to disprove in order to get rid
of Paley's argument for God. He stated the position of the opposition in such an extreme way, a way in which no creationist
known to the authors would state it, and then because that
view seemed so unreasonable, he concluded that the evolutionist

---

[6] Byron C. Nelson, *The Deluge Story in Stone,* Minneapolis:
Augsburg Publishing House.

must be right. However, what he stated was not the creationist view, but his own imaginations.[7]

The Either-Or Fallacy

Wallace, as well as Darwin, assumed that either every trivial detail was designed, or nothing was designed. This is the "all or nothing" attitude that leads to so many false conclusions. This attitude led Wallace to maintain that either there has been no change, or there has been unlimited change—such change as the hypothesis of evolution postulates. The "either-or" attitude would force one to conclude that the unlimited changes demanded by evolution had actually taken place, since some change has taken place.

From even these brief references to Wallace it is seen that he, too, had his mind prepared for evolution by his loss of faith in creation as set forth in the Bible, and by his acceptance of the theory of uniformity, as expounded by Lyell.

---

[7] See Robert E. D. Clark, *Darwin: Before and After*, pp. 88-89.

CHAPTER VII

THE ANTI-SUPERNATURAL BIAS
OF OTHER SCIENTISTS

Other scientists have been just as definite in expressing their anti-supernatural bias as were Darwin, Huxley, and Spencer. Samples of such statements from scientists will be presented now.

Henry Fairfield Osborn

One of the best known scientists in America of the past generation was Henry Fairfield Osborn. He makes a clear statement of the desire to explain everything naturally. "In truth, from the period of the earliest stages of Greek thought man has been eager to discover some natural cause of evolution, and to abandon the idea of supernatural intervention in the order of nature. Between the appearance of *The Origin of Species,* in 1859, and the present time there have been great waves of faith in one explanation and then in another: each of these waves of confidence has ended in disappointment, until finally we have reached a stage of very general scepticism. Thus the long period of observation, experiment, and reasoning which began with the French natural philosopher Buffon, one hundred and fifty years ago, ends in 1916 with the general feeling that our search for causes, far from being near completion, has only just begun."[1]

This quotation shows that it was not a study of nature itself that led men to search for some hypothesis of natural evolution but rather the desire to escape the supernatural. This evidently

---

[1] Henry Fairfield Osborn, *The Origin and Evolution of Life.* New York: Charles Scribner's Sons, 1918, pp. ix-x.

is the reason that men so eagerly grasped at any hypothesis that appeared to do just that. They did not accept a hypothesis because the evidence demanded its acceptance. This is evident from the fact that although eagerly received at first they were later abandoned. The reason that the hypotheses were accepted, and are accepted now, is that they fitted in with a frame of mind that was eager for, and demanded, just that type of explanation.

G. Fana

"[Hypotheses] on the origin of species are an indication of our mental tendencies rather than the synthetic result of facts incontrovertibly ascertained. Let us admit without further preamble: the success attained by the theory of evolution is not due primarily to its self-evident character, for even the most generally admitted facts cannot always be reconciled with it, but rather to the sympathy of the scientific world for the dogma of continuity of natural phenomena."[2]

D. M. S. Watson

In his presidential address to the Zoology Section of the British Association, Professor Watson said: "Evolution itself is accepted by zoologists not because it has been observed to occur or is supported by logically coherent arguments, but because it does fit all the facts of taxonomy, of palaeontology, and of geographical distribution, and because no alternative explanation is credible.

"Whilst the fact of evolution is accepted by every biologist, the mode in which it has occurred and the mechanism by which it has been brought about are still disputable. The only two 'theories of evolution' which have gained any general currency, those of Lamarck and of Darwin, rest on a most insecure basis; the validity of the assumptions on which they rest has seldom been seriously examined, and they do not interest most of

---

[2] G. Fana, *Brain and Heart*. Oxford University Press, 1926, p. 41. Quoted by L. M. Davies, *Transactions of the Victoria Institute*, 1929.

the younger zoologists. It is because I feel that recent advances in zoology have made possible a real investigation of these postulates that I am devoting my address to them."[3]

T. H. Huxley, however, at times stated that the idea of creation was credible. "It seemed to me then (as it does now) that 'creation,' in the ordinary sense of the word, is perfectly conceivable. I find no difficulty in imagining that, at some former period, this universe was not in existence; and that it made its appearance in six days (or instantaneously, if that is preferred), in consequence of the volition of some pre-existent Being. Then, as now, the so-called *a priori* arguments against Theism; and, given a Deity, against the possibility of creative acts, appeared to me to be devoid of reasonable foundation."[4]

Delage

Although Professor Delage, then a Professor of Comparative Zoology in the University of Paris, rejected the Darwinian theory of Natural Selection as the explanation of evolution, he "hastens to add: 'Whatever may befall this theory in the future, whether it is to be superseded by some other theory or not, Darwin's everlasting title to glory will be that he explained the seemingly marvellous adaptation of living things by the mere action of natural factors without looking to a divine intervention, without resorting to any finalist or metaphysical hypothesis.' "[5] But since he rejected natural selection, and this was Darwin's chief explanation of evolution, how could Darwin's rejected hypothesis be a claim to glory?

Weismann

" 'We must assume,' wrote Weismann, 'natural selection to be the principle of the explanation of the metamorphoses because all other apparent principles of explanation fail us, and

---

[3] D. M. S. Watson, "Adaptation," *Nature*, August 10, 1929, p. 231.

[4] *Life and Letters of Charles Darwin*, Vol. I, p. 541.

[5] Arnold Lunn, *The Flight from Reason*, p. 68.

it is inconceivable that there should be another capable of explaining the adaptation of organisms *without assuming the help of a principle of design.'*" As Arnold Lunn commented: "We must accept, so he argues, a theory which we have every reason to distrust because the only alternative implies the existence of God."[6]

Ludwig von Bertalanffy

"Behind the logical thesis that all concepts in science are reducible to physical concepts, there lies a metaphysical motive, although this would be sternly denied by the representatives of logical positivism. This motive is that the world, as pictured in physics, is the ultimate reality. The world consists of those elementary particles called atoms, electrons, protons, neutrons, and the like; and the things observed, whether stars and crystals, plants and animals, or brains and mental life, are aggregates or the outcome of those ultimate realities."[7]

William D. Matthew

Unless evolution is true these scientists would be forced to accept creation by God. This they are unwilling to do, and thus they frame whatever hypotheses are necessary to sustain a hypothesis of evolution. That this is the case is implied by a well-known evolutionist who, at the time of his death in 1930, was head of the department of geology, "and professor of paleontology and director of the paleontological museum" in the University of California.[8]

"According to some distinguished paleontologists (Deperet, Thevenin and others), progress is to be made only by ignoring

---

[6] *Ibid.*, pp. 67-68.

[7] Ludwig von Bertalanffy, *The Scientific Monthly*, November, 1953, p. 236.

[8] *Climate and Evolution*, 2nd Edition, Revised and Enlarged, Arranged by Edwin Harris Colbert, Preface by William King Gregory. June 15, 1939. Published by the Academy (N.Y.) of Sciences, pp. x-xi.

the possibility that races have originated in or migrated from regions of whose former life we have substantially no record, and assuming that they must have evolved in one or another region where the record is more or less known, and that the actual record must be the sole basis for any conclusion. They refuse to consider the arguments for origin elsewhere, on the ground that such hypotheses are 'vain speculations' and 'serve merely to conceal our ignorance.'

"To this I may answer that a fair and full consideration of the data at hand shows that such hypotheses, of one kind or another, are absolutely necessary, unless we are to abandon all belief in the actuality of evolution and are to treat it as merely a convenient arrangement of successive species and faunas independently created. Such a view was held by Agassiz and most of his predecessors, but it is unnecessary to consider it in the present state of scientific belief."[9] Such scientists, of course, would not consider it.

H. S. Shelton

In a debate with Douglas Dewar, on *Is Evolution Proved?* Shelton wrote: "I must therefore say quite bluntly that I regard the hypothesis of special creation as too foolish for serious consideration; indeed, I do not regard it as a hypothesis at all, but merely one of those peculiar confusions of thought which remove some anti-evolutionists from the class of people with whom it is possible to conduct a rational discussion."[10] Shelton has bluntly put what so many assume. Of course, he does not know—as a result of rational discussion—that the view of special creation is not right, for he has not seriously considered it, since he said that it was too foolish for serious consideration. His hasty brush-off of special creation, accompanied by his admission that it is done without serious consideration since it strikes him as foolish, is not a reflection on special creation nor on those who hold to it.

---

[9] *Ibid.*, pp. 137-138.
[10] P. 114.

Louis T. More

"The evidence for the evolution of plants and animals is commonly said to be derived from many sources. When, however, we examine these causes for our belief we find that, excepting our desire to eliminate special creation and, generally, what we call the miraculous, most of them can be considered only as secondary reasons to confirm a theory already advanced."[11]

A. L. Kroeber

Although he thought that the situation was much better today, A. L. Kroeber at the Darwin Centennial said: "Overwhelmingly, biologists had been accepting evolution because there was nothing else for them to do; but they had not proved it to their own satisfaction."[12]

Du Bois-Reymond

In eulogizing the work of Darwin, Du Bois-Reymond, a German scientist, clearly stated the anti-supernatural bias which arranged the furniture of the mind in such a way that it welcomed Darwinism as a justification of the bias. "Here is the knot, here the great difficulty that tortures the intellect which would understand the world. Whoever does not place all activity wholesale under the sway of Epicurean chance, whoever gives only his little finger to teleology, will inevitably arrive at Paley's discarded 'Natural Theology,' and so much the more necessarily, the more clearly he thinks and the more independent his judgment...the physiologist may define his science as a doctrine of the changes which take place in organisms from internal causes.... No sooner has he, so to speak, turned his back on himself than he discovers himself talking again of functions, performances, actions, and purposes of the organs. *The possibility, ever so distant, of banishing from nature its seeming pur-*

---

[11] *The Dogma of Evolution,* Princeton: Princeton University Press, 1925, p. 117.
[12] Sol Tax, Editor, *Evolution of Man,* p. 2.

pose, and putting a blind necessity everywhere in the place of final causes, appears, therefore, as one of the greatest advances in the world of thought, from which a new era will be dated in the treatment of these problems. To ·have somewhat eased the torture of the intellect which ponders over the world-problem will, as long as philosophical naturalists exist, be Charles Darwin's greatest title to glory."[13]

There are others who could be quoted, but these quotations illustrate that as in the days of Darwin, so it is now that men often accept evolution because the only alternative is creation by God.

---

[13] Du Bois-Reymond, "Darwin Versus Galiani," Vol. I, p. 216. Quoted from Merz, *History of European Thought in the Nineteenth Century.* Vol. II, p. 435, footnote.

CHAPTER VIII

EVOLUTION IS A FAITH, NOT A FACT

The title of this chapter does not imply that a faith cannot be founded on fact. Sometimes, however, there is faith that is not founded on fact, but is the creation of prejudice or credulity. In this chapter we shall show—the evolutionists being their own witnesses—that many of them grant that evolution is, after all, only a hypothesis; and that they hold to it as a faith although it has not been proved and there are many and great difficulties. Other evolutionists forget, or do not know, this and thus they write as if evolution is a scientific fact and that only the prejudiced, or the ignorant, reject it. But in the very nature of the case the hypothesis of evolution can never be anything other than a faith even though all of the arguments used to support it were sound. If the arguments of the evolutionists were sound, the hypothesis of evolution would be a reasonable faith, but it would still be held on faith without scientific demonstration.

Why Evolution Must Always Be a Faith

Evolution, if it actually took place, would now be a matter of history, and history cannot be repeated in the laboratory. By the experimental method, for example, one cannot prove that Pasteur ever worked an experiment. What experiment carried on in the laboratory today could prove that Pasteur ever lived, much less prove that he worked certain experiments? It would, of course, be possible to show this by historical testimony, and to show in experiments today that such experiments are possible. But even so, one would still have to establish by historical testimony whether they were possible in Pasteur's day and whether he actually worked experiments.

In like manner, even if it were possible in a laboratory today to create life and man, this would not prove that it happened in the past. It would prove that it is possible for intelligence to bring about such results, but it would not prove that the non-intelligent brought about evolution in the past. It would, it is true, render it quite reasonable that an intelligent being could have done it in the past. But certainly it would not prove that evolution was the product of matter in motion.

The scientist, however, cannot go into the laboratory today and demonstrate that such evolution is possible today. Neither can he go to historical testimony, as we can in the case of Pasteur. For what historian was alive and recorded the origin of life, and its manifold developments, by the process of naturalistic, or even theistic, evolution? Thus, even if the evidence for it were quite good, evolution would still be a faith and a faith it would have to remain.

Charles Darwin Knew That Evolution Had Not Been Proved

We again quote from Darwin's letter to G. Bentham (May 22, 1863): "In fact, the belief in Natural Selection must at present be grounded entirely on general considerations. (1) On its being a *vera causa*, from the struggle for existence; and the certain geological fact that species do somehow change. (2) From the analogy of change under domestication by man's selection. (3) And chiefly from this view connecting under an intelligible point of view a host of facts. When we descend to details, we can prove that no one species has changed (i.e. we cannot prove that a single species has changed): nor can we prove that the supposed changes are beneficial, which is the groundwork of the theory. Nor can we explain why some species have changed and others have not. The latter case seems to me hardly more difficult to understand precisely and in detail than the former case of supposed change."[1]

It is true that species are different, but the fossils do not show that one has developed into another by a process of nat-

---

[1] Francis Darwin, Editor, *The Life and Letters of Charles Darwin,* New York: D. Appleton and Co., 1898, Vol. II, p. 210.

ural selection or otherwise. It records differences, but not the development of one into another.

Although there are changes brought about under domestication, there have been none that prove the hypothesis of organic evolution. These limited changes have been observed but who has observed the unlimited changes demanded by the Darwinian theory?

In view of Darwin's admission, which was made after he had published the *Origin of Species*, is it not strange that many claim that his book proved the theory of organic evolution? Even he did not believe that it did. This, as we have previously remarked, is further testified to by the hundreds of "perhapses," and other statements of uncertainty, that are in the book.

The disciples of Darwin, however, do not seem to see how anyone could read his book fairly and not be convinced. And yet Darwin, in another letter to G. Bentham on June 19, 1863, said: "I, for one, can conscientiously declare that I never feel surprised at any one sticking to the belief of immutability; though I am often not a little surprised at the arguments advanced on this side. I remember too well my endless oscillations of doubt and difficulty."[2]

In writing to Herbert Spencer, on February 23, 1860, Darwin wrote: "Of my numerous (private) critics, you are almost the only one who has put the philosophy of the argument, as it seems to me, in a fair way—namely, as an hypothesis (with some innate probability, as it seems to me) which explains several groups of facts."[3]

Herbert Spencer

In the first edition of his work on psychology in 1855 Spencer, an evolutionist, wrote: "It must suffice to enunciate the belief that Life under all its forms has arisen by a progressive, unbroken evolution; and through the immediate instrumentality

---

[2] *Ibid.*, pp. 210-211.
[3] David Duncan, Editor, *Life and Letters of Herbert Spencer*, Vol. I, p. 128.

of what we call natural causes. That this is an hypothesis, I readily admit. That it may never be anything more, seems probable. That even in its most defensible shape there are serious difficulties in its way, I cheerfully acknowledge: though, considering the extreme complexity of the phenomena; the entire destruction of the earlier part of the evidence; the fragmentary and obscure character of that which remains; and the total lack of information respecting the infinitely-varied and involved causes that have been at work; it would be strange were there not such difficulties."[4]

William Berryman Scott

William Berryman Scott, once a professor of Geology and Palaeontology in Princeton University, rejected Darwinism although he accepted the hypothesis of evolution. "Personally, I have never been satisfied that Darwin's explanation is the rightful one; to one who approaches the problem from the study of fossils, the doctrine of natural selection does not appear to offer an adequate explanation of the observed facts. The doctrine, in its application to concrete cases, is vague, elastic, unconvincing and seems to leave the whole process to chance. To be sure, this difficulty is impossible; no one ever saw the birth of a species and thus we are shut up to drawing of inferences from what may be learned by comparison and experiment.

"On the other hand, if Darwin's hypothesis be rejected, there is, it must be frankly admitted, no satisfactory alternative to take its place.... In short, while the evolutionary theory is buttressed by such a mass of evidence that nearly all men of science are convinced of its truth, no satisfactory and acceptable explanation of its causation has yet been devised."[5] Scott, like most evolutionists, is sure that evolution has occurred, and thinks that there is a mass of evidence for it. But does the "mass

---

[4] Herbert Spencer, *The Principles of Psychology,* New York: D. Appleton and Co., 1897, Vol. I, pp. 465-466, footnote.

[5] *The Theory of Evolution,* New York: The Macmillan Company, 1923, pp. 25-26.

of evidence" require the explanation given to it by the evolutionist?

"While I believe that the evolutionary conception of nature is one of our permanent possessions and that it will in the future continue to direct and condition all lines of intellectual inquiry, I can understand that half a century hence the question may possibly have assumed a very different aspect."[6]

Willis and Goldschmidt

"The conclusions relating to our present problem (aside from the derivation from facts of distribution) are well expressed by Willis (1923), who says that the change from one species to another must be in one or, at most, a few large steps, changing many or all characters of the plant at once. Knowing that the geneticists will not agree, he makes the situation clear in the following words:

" 'The current attitude of the Mendelians towards questions of evolution is one of an aggressive agnosticism. Since investigations upon Mendelian lines have not as yet been able to throw as much light upon the problem as had been at one time expected, they seem to think that no other line of attack upon the question will be any more likely to find a way that may possibly lead to something in the nature of a solution of the problem at some future date. They seem inclined to think that because they have not themselves seen a "large" mutation, such a thing cannot be possible. But such a mutation need only be an event of the most extraordinary rarity to provide the world with all the species that it has ever contained. As I have pointed out (*Age and Area,* p. 212), one large and viable mutation upon any area of a few square yards of the surface of the earth, and once in perhaps fifty years, would probably suffice. The chance of seeing such a mutation occur is practically nil, whilst if the result were subsequently found it would probably be called a relic. Darwin's theory of natural selection has never had any

---

[6] *Ibid.,* pp. 171-172.

proof except from *a priori* considerations, yet it has been universally accepted, and has led to great advances in biology; and until the Mendelians show us how to control mutation (a thing that will evidently be some day possible), the proposition now put forward will presumably go without actual demonstration by verified fact. What I contend is that the facts brought up here and elsewhere go to show that neither of the extreme suppositions—Special Creation and Natural Selection—contains all the truth, and that therefore this, or similar, compromise between them is rendered necessary by the present condition of our knowledge.' "[7]

Louis T. More

"The more one studies palaeontology, the more certain one becomes that evolution is based on faith alone; exactly the same sort of faith which it is necessary to have when one encounters the great mysteries of religion. The changes that are noted as time progresses show no orderly and no consecutive evolutionary chain and, above all, they give us no clue whatever as to the cause of variations. Evolutionists would have us believe that they have photographed the successions of fauna and flora, and have arranged them on a vast moving picture film. Its slow unrolling takes millions of years. A few pictures, mostly vague, defaced and tattered, occasionally attract our attention. Between these memorials of the past are enormous lengths of films containing no pictures at all. And we cannot tell whether these parts are blanks or whether the impression has faded from sight. Is the scenario a continuous changing show or is it a succession of static events? The evidence from palaeontology is for discontinuity; only by faith and imagination is there continuity of variation."[8]

---

[7] Richard B. Goldschmidt, *The Material Basis of Evolution,* New York: Pageant Books, Inc., 1960, pp. 211-212.

[8] *The Dogma of Evolution,* Princeton, N. J.: Princeton University Press, 1925, pp. 160-161. In the authors' judgment the great mysteries of religion are not based on faith alone.

Paul Lemoine

One of the editors of the *French Encyclopaedia*, Paul Lemoine, went so far as to say that evolution was a dogma to which some continued to hold although they knew that it was not established.

"It results from this exposé that the theory of evolution is impossible. Moreover, in spite of appearance no one no longer believes it, and one says it, without attaching any importance to it otherwise, evolution in order to signify enchainment—or, more evolved, less evolved in the sense of more perfectioned, less perfectioned, because it is conventional language, admitted and almost obligatory in the scientific world. Evolution is a sort of dogma in which the priests no longer believe but that they maintain for their people.

"That—one must have the courage to say it in order that men of future generations orient their research in another way. For one must say it well also, even if the theories of evolution were truly dismounted, one would arrive at the supreme hypothesis before which all the biologists have withdrawn. How has life appeared? And why such day rather than such other? On such period rather than on such other?"[9]

"The idea of evolution is admissible for some limited groups; it is not for the masses of the animal and vegetable kingdom."[10]

T. H. Huxley

"I by no means suppose that the transmutation hypothesis is proven or anything like it."[11]

Wm. L. Straus, Jr.

"I wish to emphasize that I am under no illusion that the theory of human ancestry which I favor at the present time, can in any way be regarded as proven. It is at best merely a

---

[9] *French Encyclopaedia*, 5-82-8.
[10] *Ibid.*, 5-82-9. Translated by Constance Ford.
[11] Leonard Huxley, *Life and Letters of T. H. Huxley*, Vol. I, p. 252, June 25, 1859.

working hypothesis whose final evaluation must be left to the future. What I am trying to point out is that, from what we now know, this interpretation appears to be distinctly more valid than the orthodox, anthropoid-ape theory. The ultimate verdict, if there can be a final verdict in such a matter, will rest upon paleontological evidence at present lacking; for with due respect to the Australopithecinae, the gap in the fossil record between man and the other primates remains very great indeed.

"What I wish especially to stress is that the problem of man's ancestry is still a decidedly open one, in truth, a riddle. Hence it ill behooves us to accept any premature verdict as final and so to prejudice analysis and interpretation of whatever paleontological material may come to light as the orthodox theory has so often done and is still doing. One cannot assume that man is made-over anthropoid ape of any sort, for much of the available evidence is strongly against that assumption."[12]

J. M. Gillette

"It is evident that anthropologists assume and profess to be evolutionists. They probably would resent the suggestion that they are dogmatists in the biological field, yet such appears to be the case regarding the subject under discussion. By their works ye shall know them, and it is by their work products in this particular field of evolution that they are now to be judged. Their genealogical trees purporting to show the evolution of man should satisfy theological fundamentalists who reject the idea of such evolution. In fact they are empty forms which consist of nothing but assumed roots, trunk, many limbs which grow in number through the years, and human twigs terminating the trunk which are supposed to connect with the assumed roots. Now a tree that is constituted wholly of limbs does not tell us much. Limbs below do not beget the limbs above them. They are not ancestral to them, only cousins to what is above them.

"Biologists are, of course, confessedly evolutionists, but it is

---

[12] *The Quarterly Review of Biology,* Sept. 1949, p. 220.

really remarkable how little evidence they admit in support of their position."[13]

William Bateson

This authority on genetics, of the past generation, considered evolution as a faith. "When students of other sciences ask us what is now currently believed about the origin of species we have no clear answer to give. Faith has given place to agnosticism for reasons which on such an occasion as this we may profitably consider."[14] He was convinced that evolution took place, but he was utterly in the dark as to how. "Meanwhile, though our faith in evolution stands unshaken, we have no acceptable account of the origin of 'species.'"[15]

Thompson

"There is one last lesson which coordinate geometry helps us to learn; it is simple and easy, but very important indeed. In the study of evolution, and in all attempts to trace the descent of the animal kingdom, fourscore years' study of the *Origin of Species* has been an unlooked-for and disappointing result. It was hoped to begin with, and within my own recollection it was confidently believed, that the broad lines of descent, the relation of the main branches to one another and to the trunk of the tree, would soon be settled, and the lesser ramifications would be unravelled bit by bit and later on. But things have turned out otherwise. We have long known, in more or less satisfactory detail the pedigree of horses, elephants, turtles, crocodiles and some few more; and our conclusions tally as to these, again more or less to our satisfaction, with the direct evidence of palaeontological succession.

"But the larger and at first sight simpler questions remain

---

[13] "Ancestorless Man: The Anthropological Dilemma," *Scientific Monthly,* Dec. 1943, p. 533.

[14] Toronto Address, 1922, Beatrice Bateson, *William Bateson, F.R.S. Naturalist. His Essays and Addresses Together with a Short Account of His Life,* Cambridge: University Press, 1928, p. 391.

[15] *Ibid.,* pp. 394, 398.

unanswered; for eighty years' study of Darwinian evolution has not taught us how birds descend from reptiles, mammals from earlier quadrupeds, quadrupeds from fishes, nor vertebrates from the invertebrate stock. The invertebrates themselves involve the selfsame difficulties, so that we do not know the origin of the echinoderms, of the molluscs, of the coelenterates, nor of one group of protozoa from another. The difficulty is not always quite the same. We may fail to find the actual links between the vertebrate groups, but yet their resemblance and their relationship, real though indefinable, are plain to see; there are gaps between the groups, but we can see, so to speak, across the gap. On the other hand, the breach between the vertebrate and invertebrate, worm and coelenterate, coelenterate and protozoan, is in each case of another order, and is so wide that we cannot see across the intervening gap at all."[16]

Although we do not agree with every statement in this quotation from Dr. Thompson, he does show how their faith in evolution has led them in times past to lean on "broken reeds" that were entirely inadequate. He shows how they constructed hypotheses—such as the emphasis on the extreme imperfection of the geological record—to explain away the impact of facts that did not harmonize with the picture which evolutionists constructed.

These sudden changes *must* have taken place, Dr. Thompson argues. But why *must*? Because he has already accepted as true the hypothesis of evolution, and since there is no other way to bridge the gaps it is *certain* that they were bridged by mutations either large or small. Yes, it is certain because evolution must be true! The authors agree with Dr. Thompson that there are such gaps, but they do not agree that these have been bridged by natural means, or that there is any proof that they have been so bridged. And with reference to many of these gaps, Dr. Thompson admits that eighty years of study of Darwinism has not instructed us as to how it could be. Perhaps Dr. Thompson

---

[16] D'Arcy Wentworth Thompson, *On Growth and Form,* Cambridge: University Press, 1943, new edition, pp. 1092-1093.

believed that God had something to do with it, as toward the close of the book he stated his faith in God.

It is not the purpose of this book, however, to examine the specific arguments for or against the theories of organic evolution, but to show that it was not accepted for scientific reasons. It is the purpose of this chapter to show that some leading scientists acknowledge that the theories have not been scientifically established, but that they are held by faith. With Professor Emanuel Radl, Professor of Natural Philosophy in the University of Prague, they grant that "It is true that the theory has not received any clinching scientific proof." He, however, believed that it was very probable and that objections to it were slight; and that, therefore, we should safely assume its truth.[17]

---

[17] *The History of Biological Theories*. Translated and Adapted from the German by E. J. Hatfield. Oxford University Press, 1930, pp. v-vi.

CHAPTER IX
THE CONCLUSION

Evolution is taken for granted today and thus it is uncritically accepted by scientists as well as by laymen. It is accepted by them today because it was already accepted by others who went on before them and under whose direction they obtained their education. They are not only brought up to believe it, but, as Thomas Dwight, then Parkman Professor of Anatomy at Harvard, observed, it is heresy to reject it. "The tyranny of the *Zeitgeist* in the matter of evolution is overwhelming to a degree of which outsiders have no idea; not only does it influence (as I must admit that it does in my own case) our manners of thinking, but there is the oppression as in the days of the 'terror.' How very few of the leaders of science dare tell the truth concerning their own state of mind! How many feel themselves forced in public to do a lip service to a cult they do not believe in! As Professor T. H. Morgan intimates, it is only too true that many of these who would on no account be guilty of an act which they recognize as dishonest, nevertheless speak and write habitually as if evolution were an absolute certainty as well established as the law of gravitation."[1]

Professor Paul Shorey, then of the University of Chicago, stated that "An ambitious young professor may safely assail Christianity or the Constitution of the United States or George Washington or female chastity or marriage or private property. ... But he must not apologize for Bryan.... It is not done."[2]

---

[1] *Thoughts of a Catholic Anatomist,* London: Longmans, Green and Co., 1927, pp. 20-21.
[2] *Atlantic Monthly,* Vol. 142, Oct., 1928, p. 478.

The tendency to conformity is so great that Solomon E. Asch pointed out that there are many people who will call white black in order to be in step with the times. "Life in society requires consensus as an indispensable condition. But consensus, to be productive, requires that each individual contribute independently out of his experience and insight. When consensus comes under the dominance of conformity, the social process is polluted and the individual at the same time surrenders the powers on which his functioning as a feeling and thinking being depends. That we have found the tendency to conformity in our society so strong that reasonably intelligent and well-meaning young people are willing to call white black is a matter of concern. It raises questions about our ways of education and about the values that guide our conduct."[3]

Although one should not be a non-conformist just for the sake of being a non-conformist, yet the pressure to accept evolution is so strong that there are many who accept it because they are afraid of what others may think. All, however, realize that a thing is not true just because it is already accepted, or because there is pressure exercised in order to insure that people will continue to accept it. Otherwise, one would have to argue that every idea which is accepted by a number of people is true; and that, therefore, it is unreasonable for a person in any group to question any idea that is held by the group which educated him.

It is also well known that false hypotheses may be accepted and passed on without being critically examined. The hypothesis itself may not be seriously questioned; instead men may spend their time trying to find proof for it. As one well-known evolutionist, and one-time head of the Department of Geology in the University of California, put it: "Many a *false* theory gets crystallized by time and absorbed into the body of scientific doctrine through lack of adequate criticism when it is formulated."[4]

---

[3] *Scientific American*, Nov., 1955, p. 34.
[4] William Diller Matthew, *Climate and Evolution*, 2nd Edition, p. 159.

Because a few eminent men first accepted the hypothesis of organic evolution, the multitudes—in science and out of science —accept it today. These do not accept it for the same reasons, whether sound or unsound, that these eminent men accepted it. They accept it for the simple reason that certain men, who were supposed to know, accepted it. Because these men were outstanding, and intelligent, multitudes conclude that they would not have accepted evolution if there had not been an abundance of convincing evidence to justify their so doing. This is not merely the assumption of the non-scientists, but of many scientists. The fact that an individual is a scientist does not mean that he is well acquainted with the history of science in general, or of the history of his field of science in particular. Much less does it mean that he is acquainted with the reasons why scientists in the nineteenth century accepted evolution. This is underscored further by the fact that the reason these men accepted evolution is not brought out clearly in their scientific works but in their letters, biographies and autobiographies which many scientists have never examined. And yet, these materials furnish evidence that is very pertinent to the issue under consideration.

When, however, it becomes generally known that the eminent evolutionists of the nineteenth century accepted evolution because of their anti-supernatural bias, and not because of the weight of scientific evidence, at least two conclusions will logically follow. First, that an individual is not necessarily either ignorant or dishonest because he rejects the hypothesis of evolution concerning life's origin and manifold forms. Thus evolutionists ought to be more tolerant of those who do not accept evolution. Second, that the question of the truth of evolution itself should be reopened. At the present the truth of evolution is assumed. The research of certain scientists is not directed toward determining whether or not evolution has taken place but, accepting the idea that it has, their efforts have been directed toward proving *how* that which must have taken place did take place. A study of whether or not evolution has been scientifically established could lead to a third conclusion. It could lead more scientists to take a scientific attitude toward evolution and to accept the position that evolution is simply a working hypothesis, which thus should be treated as a working hypothesis and not as a fact. Why is it that such an attitude is taken toward other matters which are no more scientifically established than is evolution, but that it is heresy to take the same attitude toward evolution? The acceptance of evolution as simply a hypothesis could lead to the preparation of textbooks wherein scientists who are creationists and scientists who are evolutionists presented both sides of the question of origins.

In the light of the documentation presented in this book it is not unscientific to suggest that the question of the truth or falsity of evolution be reopened; and that the hypothesis of evolution be required to pass as rigid tests as are required for other hypotheses before they are considered to be theories, and before they are viewed as laws. In other words, the authors are asking scientists to be scientific in their treatment of the hypothesis of evolution. Then, instead of ascribing unknown effects to known causes, they will at least admit as is done in a widely used textbook in biology that "The piecing together of the evolution story is comparable to the reconstruction of an atom-bombed metropolitan telephone exchange by a child who has only seen a few telephone receivers. We know something about living plants and animals, and we have some fossil remnants to go on. Extensive study of the evidence available plus ingenious hypotheses, most of which cannot be adequately tested, have given us a sort of a trial schedule of the possible directions of evolution of living organisms."[5]

---

[5] Relis B. Brown, *Biology*, second edition, Boston: D. C. Heath and Company, 1961, p. 531.

BIBLIOGRAPHY

1. ADAMS, FRANK DAWSON. *The Birth and Development of the Geological Sciences.* Baltimore: The Williams and Wilkins Company, 1938.

2. ARTHUR, WILLIAM. *Religion without God.* London: Bemrose and Sons, 1888.

3. ASCH, SOLOMON E. *Scientific American,* Nov., 1955, p. 34.

4. BALES, JAMES D. *Communism: Its Faith and Fallacies.* Grand Rapids 6, Michigan: Baker Book House, 1962.

5. BALLARD, FRANK. *The Miracles of Unbelief.* Edinburgh: T. & T. Clark, 38 George Street, 4th Edition, 1902.

6. BATESON, BEATRICE. *William Bateson, F.R.S. Naturalist.* Cambridge: At the University Press, 1928.

7. BERTALANFFY, LUDWIG VON. *The Scientific Monthly,* November, 1953, p. 236.

8. BROWN, RELIS B. *Biology,* 2nd Edition. Boston: D. C. Heath & Co.

9. BRUCE, F. F. *Are the New Testament Documents Reliable?* London: The Inter-Varsity Fellowship, 1948.

10. CLARK, HAROLD W. *The New Diluvianism.* Angwin, California: Science Publications, 1946.

11. CLARK, ROBERT E. D. *Conscious and Unconscious Sin.* London: Williams and Norgate, 1934.

12. ———. *Darwin: Before and After.* London: The Paternoster Press, 1948.

13. DARWIN, CHARLES. *Descent of Man,* Vols. I, II. New York: P. F. Collier & Son, 1905.

110

14. DARWIN, FRANCIS (ED.). *Life and Letters of Charles Darwin.* New York: D. Appleton and Co., 1898, Vols. I, II.

15. ———. *More Letters of Charles Darwin.* New York: D. Appleton and Co., Vols. I, II.

16. DAVIES, L. M. *Transactions of the Victoria Institute.* London: Philosophical Society of Great Britain, 1929.

17. DEWAR, DOUGLAS. *More Difficulties of the Evolution Theory.* London: Thynne and Co., Ltd., 1938.

18. ———. *The Transformist Illusion.* Murfreesboro, Tennessee: DeHoff Publishing Co., 1957.

19. DUNCAN, DAVID. *Life and Letters of Herbert Spencer.* New York: D. Appleton and Co., 1908, Vols. I, II.

20. DWIGHT, THOMAS. *Thoughts of a Catholic Anatomist.* London: Longmans, Green, and Co., 1927.

21. *Evolution,* Toronto, Ontario, Canada: International Christian Crusade.

22. FITCHETT, W. H. *The Beliefs of Unbelief.* New York: Cassell Co., Ltd., 1911.

23. FREDERICK, MARY. *Religion and Evolution Since 1859.* Notre Dame: University of Notre Dame, 1934.

24. *French Encyclopaedia,* 5-82-8; 5-82-9.

25. GILLETTE, J. M. "Ancestorless Man: The Anthropological Dilemma," *Scientific Monthly,* Dec., 1943.

26. GOLDSCHMIDT, RICHARD B. "Evolution, As Viewed by One Geneticist," *American Scientist,* Vol. 40, 1952.

27. ———. *The Material Basis of Evolution.*

28. HAMILTON, FLOYD. *The Basis of Christian Faith.* New York: Harper and Co.

29. HATFIELD, E. J. (translator and adapter). *The History of Biological Theories.* Oxford University Press, 1930.

30. HOFSTADTER, R. *Social Darwinism.*

31. HUTTON, JAMES. "Theory of the Earth," *Transactions of the Royal Society.* Edinburgh, 1785.

112

32. HUXLEY, JULIAN S. *Evolution: The Modern Synthesis*. New York: Harper and Brothers, 1943.

33. HUXLEY, LEONARD. *Life and Letters of T. H. Huxley*, Vols. I, II, III. London: Macmillan and Co., Ltd., 1903.

34. HUXLEY, T. H. *Darwiniana*. New York: D. Appleton and Company, 1896.

35. ———. *Evolution and Ethics and Other Essays*. New York: D. Appleton and Co., 1896.

36. LEWIS, C. S. *Miracles: A Preliminary Study*. New York: The Macmillan Co., 1959.

37. LUNN, ARNOLD. *The Flight from Reason*. London: Eyre & Spottiswoode, Ltd., 1931.

38. ———. *The Revolt Against Reason*. New York: Sheed and Ward, 1951.

39. LYELL, MRS. *Life, Letters, and Journals of Sir Charles Lyell*. London: John Murray, 1881, Vols. I, II.

40. MARCHANT, JAMES. *Alfred Russell Wallace: Letters and Reminiscences*. London: Cassell and Co., Ltd., 1916, Vol. I.

41. MATHER, KIRTLEY F. AND MASON, SHIRLEY L. *A Source Book of Geology*. New York: McGraw-Hill Book Company, Inc., 1939.

42. MATTHEWS, WILLIAM DILLER. *Climate and Evolution*, 2nd Edition. New York: New York Academy of Science.

43. McGARVEY, J. W. *Evidences of Christianity*. Cincinnati: Standard Publishing Co., 1886.

44. MERRELL, DAVID J. *Evolution and Genetics*. New York: Holt, Rinehart and Winston, 1962.

45. MERZ. *History of European Thought in the Nineteenth Century*. Vol. II.

46. MORE, LOUIS T. *The Dogma of Evolution*. Princeton: Princeton University Press, 1925.

47. NELSON, BYRON C. *The Deluge Story in Stone*. Minneapolis: Augsburg Publishing House.

48. Osborn, Henry Fairfield. *The Origin and Evolution of Life*. New York: Charles Scribner's Sons, 1918.
49. Shorey, Paul. *Atlantic Monthly*, Vol. 142, Oct., 1928.
50. Spencer, Herbert. *Autobiography*. New York: D. Appleton & Co., 1904, Vols. I, II.
51. ———. *First Principles*. New York: D. Appleton & Co., 1897, 4th Edition.
52. ———. *The Principles of Psychology*. New York: D. Appleton & Co., 1897, Vol. I.
53. Straus, William L., Jr. *The Quarterly Review of Biology*, Sept., 1949, p. 220.
54. Tax, Sol (Ed.). *Evolution of Man*. Chicago: University of Chicago Press, 1960.
55. Thompson, D'Arcy Wentworth. *On Growth and Form*. Cambridge: University Press, 1943, New Edition.
56. Ward, Henshaw. *Charles Darwin and the Theory of Evolution*. New York: The New Home Library, 1943.
57. Warfield, B. B. *Studies in Theology*. Philadelphia: The Presbyterian and Reformed Publishing Co., 1932.
58. Watson, D. M. S. "Adaptation," *Nature*, August 10, 1929.
59. Wells, H. G. *Mind at the End of Its Tether*. New York: Didier Publishers, 660 Madison Avenue, 1946.
60. Young, C. W., et al. *The Human Organism and the World of Life*. New York: Harper and Brothers, 1938.